Vacation Camping

Jeannette Augustus Marks

Alpha Editions

This edition published in 2024

ISBN : 9789362097088

Design and Setting By
Alpha Editions
www.alphaedis.com
Email - info@alphaedis.com

Contents

Contents

CHAPTER I
CAMPING CHECK LISTS

There are some considerations in camping which are staple; that is, questions and needs all of us have to meet, just as there are staple foods which all of us must have. No one knows better than the old camper, who has shaken down his ideas, theories, practices, year after year in the experiment of camping how true this is. If one is wise, one goes well prepared even into the simple life of the woods or mountains or lakes; and it is in a practical way, and under three so-called check lists, (1) camp clothes, (2) camp food, and (3) camp equipment, that I wish to tell you something about camp life for girls.

From the point of view of clothes there are two kinds of camping: one more or less civilized, the other "rough." In the first perhaps we shall be allowed a small box or trunk. In the second we have to depend entirely upon a duffle bag or a knapsack. To the camper who plans for a good many comforts, there is only one warning to be given: don't be foolish and take finery of any sort with you. Not only will it be in the way, but also a girl does not look well in the woods dressed in clothes that belong to the home life of town or city.

There is an appropriate garb for the wilderness even as there is the right gown for an afternoon tea. Except for this warning, what you will put in your trunk will be simply an extension of the comforts which you have in duffle bag or knapsack.

As the capacity of duffle bag or knapsack is very limited, the check lists for its contents must be made out with rigid economy. The most important item is foot gear. A well-made pair of medium weight boots, carefully tanned, drenched with mutton tallow, viscol, neat's-foot oil, or some similar waterproof substance, will prove the best for all-round usefulness. These boots must be broken in or worn before the camping expedition is undertaken. Nothing is so foolish as to start out in a new pair. Have in addition to the boots a pair of soft indoor moccasins. These are good to loaf around camp in. They are grateful to tired feet, and, rolled, take up but little space in the knapsack. To the boots and moccasins add from two to four pairs of hole-proof stockings of some reliable make. If you can get a really first-class stocking and are crowded for space, two pairs will do. One goes on to your feet and the other into your knapsack. There should also be several combination suits, preferably of two weights, high necked, and with shoulder and knee caps.

Now, see that the skirt you wear is of durable material; blue serge or tweed (corduroy is often too heavy); that it has been thoroughly shrunk, and is six inches off the ground anyway. Twelve would be better. Your skirt should be provided with ample pockets; the sweater and jacket also. Under the skirt wear a pair of bloomers, the lighter and slimsier they are, the better; and the stouter the material, the more practical for wear. I have tried many kinds, and believe percaline which is light, strong, slimsy and washable, the best. Silk is not suitable at all. A flannel shirt waist or blouse, a windsor or string tie, a soft felt hat with a sufficiently wide brim, but not too wide, complete your costume.

Into the knapsack put two coarse handkerchiefs, a silk neckerchief to tie around your neck, the stockings and combination suit already mentioned, a string of safety pins clipped one into another, a toothbrush, tubes of cold cream and tooth paste (tubes take up the least room and are the easiest to carry), a cotton shirtwaist, a nail file, comb, small bottle of the best cascara sagrada tablets, a pair of cotton gloves for rough work, a cake of castile soap, a towel, a stiff nail brush, *and, if you are wise,* a book for leisure hours, preferably an anthology of poems or a collection of essays which will afford food for reflection.

In your preparations let it be the rule to strip away every unnecessary article. Take pride in getting your kit down to the absolute minimum. Keep weeding out what you don't need, and then after that, weed out again.

The same principle of rigid economy in selection will obtain in the check list for food. It is the minimum of expense in the woods that will bring the maximum of comfort. In arranging for the "duffle" to be taken with you there is one thing that can be counted upon with mathematical certainty: hunger. You are going to be hungrier than you have been in a long time. The problem is, then, how to tote enough food and *get* enough food to supply your wants. The carriage, the keeping, the nutritive value, all these things have to be taken into consideration in wood life. At home we have fresh vegetables, fresh fruits, fresh meats in abundance. How can we supply these things for our camp table? We can't! But desiccated potatoes, dried apples, apricots, prunes, peaches, white and yellow-eye beans, dried lima beans, peas, whole or split, onions, rice, raisins, nuts, white and graham flour, corn meal, pilot biscuit, rolled oats, cream of wheat, cocoa (leave coffee and tea at home), sweet chocolate, syrup for flapjacks, baking soda, sugar, salt, a few candles (helpful for lighting a fire in wet weather, as well as good for illumination), matches, molasses, a little olive oil—all these things, with careful planning, we may have in abundance. To these items you should add good butter—the best salted butter is none too good—some cans of condensed milk and evaporated milk and cream, and a flitch of bacon. Meat makes a dirty camp, and a dirty camp means skunks and hedgehogs prowling around. In a properly thought-out dietary it will be entirely unnecessary to tote meat. All that is needed for use you can get at the end of your fish rod or through the barrel of your shotgun, and upon the freshness of what you catch or shoot you can depend. Dr. Breck, in his "Way of the Woods," says that if he were obliged to choose between bacon and dried apples and chocolate, he would always take the apples and chocolate. Both portage and health will be served by avoiding the carriage of a lot of tin cans. The ration of each article needed you can work out with your mother or housekeeper, according to the number of people to be in the party, the menus you plan, and the length of your stay. For a cooler for your food, you will find a wire bait box, sunk in clean running water, excellent. The question of grub, or duffle, as it is called in camp life, in proper variety, abundance and freshness, is the most difficult question of all. To this problem a seasoned camper will give his closest attention.

There are other articles, plus the food stuffs, which we must add to our check lists—chiefly articles of equipment. Two or three pails nesting into each other, a tin reflector baker for outdoor cooking, enamel-ware plates, cups and bowls, pans, dishpans, dishmop, chain pot-cleaner, double boiler, broiler, knives, forks, spoons, pepper and salt shakers, flour sifter, rotary can opener, long-handled and short-handled fry pans, a carving knife and a fish knife. The cost of these things carefully bought, will be about six dollars. There should also be in your kit some nails and a hatchet, toilet paper, woolen blankets, mosquito netting (tarlatan is better), twine, tacks, oilcloth for camp table, and some fly dope.

With these articles, plus a little knowledge of woodcraft, there is almost no wilderness into which a capable girl cannot go and make an attractive home. But a little woodcraft we must know; the rest we can learn as we go. There is one fuel in the woods which skillfully used will kindle any fire, even a wet fire, and that is birch bark. You can always get an inner layer of dry birch bark from a tree. Keep a check list of different kinds of wood and have it handy until you learn these woods for yourself. Brush tops or slashings will help to start a quick blaze. Hickory is fine for a quiet hot fire. The green woods which burn readily are white and black birch, ash, oak and hard maple. Look for pitch, which you are most likely to find in old trees, and that will always help out and start any fire. Woods that snap, such as hemlock, spruce, cedar and larch, are not to be recommended for camp fires, as a rule. To be careless or stupid about the camp fire may be to endanger the lives not only of thousands of wild creatures in the wilderness, but also the lives of human beings.

Be careful to have pure water to drink. You cannot be too careful. If you are in doubt about the water, don't drink it, or at least not until it has been thoroughly boiled. Take with you, besides those I give, a few useful recipes for cooking experiments. They will bring pleasure and variety on dull days. Choose a good place for your cabin or shack or tent, whichever you use, especially a place where the natural drainage is good. Know before you set out whether black flies, mosquitoes and midges have to be encountered and go prepared to meet them. They are sure to meet you more than halfway. Don't take any risks on land or water. The people who know the way of the woods best are those who are least foolhardy. Common sense is the law that reigns in the wilderness, and, in having our good time, we cannot do better than to follow that law.

So much for skeleton check lists, many of which, in the chapters to come, at the cost of repetition, I shall amplify. Among the questions which I shall take up are the all-important ones of camp clothes, camp food, cooking, the place, camp fires, furnishing the camp, the pocketbook, the camp dog, the outdoor training school, the camp habit, wood culture, camp health, camp friendship, homemade camping, the canoe, fishing, and the trail. This great, big, beautiful country of ours is full of girls, real CAMP FIRE GIRLS, who love the keen air of out of doors and the smell of wood smoke and the freedom of hill and lake and plain, and to them I want my little book to come home and to be a camp manual which will go with them on all journeys into the wilderness.

———————

CHAPTER II
CAMP CLOTHES

If you have been camping once, there is no need for any one to help you decide what wearing apparel to take the next time. Through the mistakes made and the discomforts involved, the girl will have learned her lesson too well to forget it. But there is always the girl who has not been camping. It is chiefly for her benefit that I am writing these chapters on camp life for girls.

In the first place, there are two kinds of camp clothes to be considered, for there are two kinds of camping: (1) the expedition which permits taking a box or trunk with you, and (2) the rougher camping that allows only the carrying of a duffle bag or a knapsack. If you are limited to a knapsack or a duffle bag, your kit must be of the most concentrated sort and chosen with the greatest care. You will find ten or fifteen pounds the most you wish to tote long distances, although at the beginning this size of pack may seem like nothing at all to you. As I have found personally, even seven pounds, with day after day of tramping, may make an unaccustomed shoulder ache under the strap.

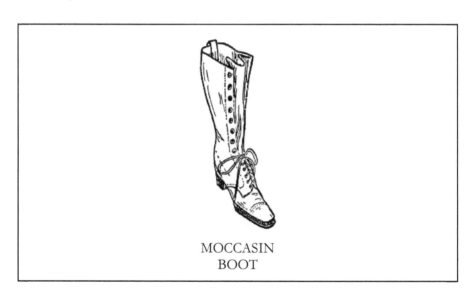

MOCCASIN
BOOT

TOBIQUE MOCCASIN

HURON INDIAN
MOCCASINS

MOCCASIN SHOE

MECCOMOC OXFORD

ELKSKIN MOCCASIN

If you are to be limited to a small duffle bag, or a fairly capacious knapsack, what are the articles of clothing without which no girl can start? Let us take up the most important item first, and that is foot-gear. Wear a well-made pair of medium weight boots, thoroughly tanned, soaked with viscol, or rubbed with mutton tallow both on the inside and the outside, to make them waterproof. *Never start out with a new pair of boots on your feet.* If necessary, get your boots weeks beforehand, and wear them from time to time till they are thoroughly
 comfortable. In addition to these boots which you wear, take a soft pair of indoor moccasins. These can be worn when you are tired and loafing around camp, or while the guide is drying or greasing your boots. If you have ever worn moccasins and are going to tramp in a moccasin country, that is, a country of forest trails and ponds, then buy a pair of heavy outdoor moccasins; larrigans or ankle-moccasins are best. These should not be too snug. Worn over a heavy cotton stocking, or a light woolen one, or woolen stockings drawn over cotton, the moccasin is the most ideal foot-gear the wilderness world can ever know.[1] Neat's-foot oil is also excellent for greasing moccasins. Buy from two to four pairs of hole-proof stockings of some reliable make. If these stockings are first class and can be depended upon, two pairs will do. One pair you will wear, the other goes into your knapsack. Have also several combination suits, some for your bag and one for your back. These suits should be high-necked and with shoulder and knee caps; of sufficient warmth for cold days and nights; in any case porous and of two weights.

[1] If you have room take with you an extra pair of shoes. When you have become a real woodswoman you will never be without woolen socks and moccasins. The thick, soft sole of sock and moccasin spare tender feet which are not accustomed to hard tramping and rough paths.

If you are going to tramp in a skirt, as you must if your route touches upon civilization, *see that it is short.* Six inches off the ground is none too much, and twelve is a good deal better. In an outing of this sort it is as poor form to wear a long skirt as it would be to wear a short skirt at an afternoon tea in civilization. The skirt should be of some good quality khaki, army preferably, or a tweed; it should be thoroughly shrunk, and if it seems desirable, it should be possible to put this camp skirt in water and wash it.[2] Have ample pockets on either side of the front seams. If I had to choose between the best of sweaters and a jacket with a lot of pockets in it, I should always choose the latter, and that is not on account of the pockets alone, but because it is a more convenient article of clothing. In case of cold weather it affords better protection, also better protection against rain as well as cold. You can have it made with two outside pockets and several inside—the more the merrier. Underneath the skirt wear a pair of bloomers. The lighter and stouter these are, the more of a comfort they will be. I have found a good quality of percaline to be the best investment. Percaline is light, strong, slimsy after a little wearing, and washes well. I have never yet found a silk that was practicable in the woods. Silk bloomers go well with the comforts of civilization, but they are not fit to endure the test of roughing it. A flannel shirtwaist or blouse, a Windsor or string tie, a soft felt hat—you may have it as pretty as you wish, provided it is not too large or over trimmed—complete the outfit which you carry on you, so to speak.

[2] You can buy an ideal hunting suit at any of the big shops in Boston, New York or Chicago for from $8 to $10.

Now to return to the outfit you carry in your pack and not on your back. A pair of indoor moccasins, an extra pair of hole-proof stockings (these you must have, not only on account of a possible wetting, but also because the stockings must be changed every day, for you cannot take too good care of your feet), two coarse handkerchiefs of ample size, a silk neckerchief to tie around your neck, an extra combination suit, a few safety pins clipped one into another until you have made a string of them, a tooth brush, a little tube of cold cream and a tube of tooth paste (the tubes are not breakable and take up the least room, they are therefore the best to carry), a cotton or linen shirtwaist of some kind, a nail file, a comb, a small vial of cascara sagrada tablets, several rolls of film for your camera—the camera itself can be slung on a strap from the knapsack—a pair of garden gloves for rough work with sooty pots and kettles, a good-sized cake of the best castile soap, a towel, a good stiff nail brush, and one or two books.

Personally I feel that the books are as indispensable as anything in the knapsack, for in moments of weariness, or when storm-bound, they prove the greatest comfort and resource. The volume taken must not be a novel which read through once one does not care to read again. Better to take some book over which you can or must linger. I have tramped scores of miles with the "Oxford Book of English Verse" in my knapsack, and it has proved the greatest imaginable pleasure and solace. A small anthology or a book of essays, or something that you wish to study, as, for example, guides about the birds or the trees or the flowers, are good sorts of volumes to tote with you—besides, of course, this camping manual.

Your kit for the rougher kind of camping, provided you have guides or men folks who will carry the food, or "grub," as it is called in camp parlance, and the blankets, is now complete. But for the one girl who goes on this rougher sort of camping expedition, twenty go into the woods to be happy in a quite civilized log cabin or shanty. These girls will be taking a camp box with them, or a trunk, and can add to their wardrobe. There is no excuse, however, for adding the wrong sort of thing. There is no excuse for wearing unsuitable, unattractive old rags about camp, clothes which have served their civilized purpose and have no fitness for the wilderness life. Let me give you one other word, from an old timer at camping, about what you should wear. *Don't be foolish and put in any finery.* The finery is as out of place in camp as your camp boots would be at a garden party at home. But several middy blouses, more shoes, more stockings, another skirt, a number of towels, a few more books—all will prove just that much added food for pleasure; first, last, and always, be comfortable in camp. There is no reason for being uncomfortable unless you enjoy discomfort. Anything, however, over and above what you actually need will be only a hindrance. Those who go camping, if they go in the right spirit, are looking for the simple life; they want to get rid of paraphernalia, not to add to it. To learn the happy art of living close to nature, means stripping away unnecessary things. There is no place in camp life for fussiness or display of any sort. All that is beyond the daily need is so much litter and clutter, making of camp life something that is a burden, something that is untidy, uncomfortable, confused. Of no thing is this more true than of a girl's camp clothes.

———

CHAPTER III
FOOD

There are several reasons why the camp food is almost more important than any other consideration. To begin with, most girls are leading a more active life than they are accustomed to living at home. This makes them hungry, and, add to the exercise the natural tonic of invigorating air, the camper becomes fairly ravenous at meal time. There are other reasons, too, why food is an all-important question. If one is in the real wilderness, it will be difficult to get. One is obliged, therefore, to consider carefully beforehand the kinds of food necessary for a well-provided table and a well-balanced diet. Another reason for taking thought about this whole subject is the portage. All the foods must be toted in, and not all kinds will prove suitable or economical in the long run for this sort of portage. Finally, there is the question of the ways and means for keeping the food, after it is once safely in camp, in good condition.

As a rule, when we go on our expeditions we leave regions where it is easy to get a great variety of foods. The city or its suburb or a comfortable country town, is the place we call home. Our tables are filled the year long with fresh vegetables, fresh fruits, fresh meats, and all kinds of bread. This dietary in all its variety, to which we have been accustomed at home, is quite impossible of realization in the camp. We might just as well make up our minds to that at once. Yet accustomed to vegetables and fruits as we are, we need them both in wholesome quantities. How shall we get them? Potatoes of course, if the camping expedition is for any length of time, that is ten days or more, must be lugged. And lugging potatoes is heavy work over a trail. As for the other vegetables and fruits, and even meats, most people buy large quantities of tinned articles and so get rid of the whole question. Personally I think that this is a great mistake. It was a delight to me to find in Doctor Breck's "Way of the Woods" that he, if obliged to choose between bacon and dried apples and chocolate, would always choose the chocolate and dried apples. And when the question of portage as well as health enters in, it may be said right here that it is quite impossible to carry a pack full of tins. But aside from the comfort of the guides, a tin-can camp is not likely to be a wholesome one. I am convinced that tin-can camping is responsible for whatever ills people experience when they go into the woods.

It is quite simple to get different kinds of dried vegetables and different kinds of dried fruits—and the best are none too good—in bulk. At present there are even evaporated potatoes on the market for campers. Such dried foods pack and carry best and are most wholesome. Both white and yellow eye beans, dried lima beans, peas, whole and split, onions, evaporated apples, dried prunes, dried peaches and apricots, rice, raisins, nuts of all kinds, lemons, oranges, and even bananas, if they are sufficiently green, can be quite easily taken into camp. Various sorts of flour and meal, too, will be needed. Find out how much it takes to bake the bread at home and add that to the length of your stay plus the number of the campers and plus a little more than you actually need, and you will be able to work out the flour problem for yourselves. There should be then white and graham flour, or entire wheat, corn meal, pilot bread (memories of toasted pilot bread in camp can make one smile from recollected joy), some rolled oats, cereals like cream of wheat which carries well, cooks easily, and is hearty, and various sorts of crackers.

Now the writer does not think meat necessary in camp. Except for the fish caught and the birds shot, none need be eaten. All the meat element or proteid necessary is provided for in the beans, peas, and nuts. But it is well to take a flitch of bacon or a few jars of it to use in broiling or frying the fish or game. Pork and lard are entirely uncalled-for in a properly thought out dietary.[3] Sufficient good fresh butter is very much needed. If campers feel that they must have other tinned meats, the best kinds to take are the most expensive, ox tongue, and that sort of thing. Several months ago four of us started off on a ten days' camping expedition into a very northern wilderness unknown to us. One of the party, needlessly ambitious, took a preserved chicken in a glass jar bought from the finest provision house in Boston. By the time we reached our destination, the chicken was anything but preserved. Indeed, unless all signs failed, it had already embarked upon a new incarnation. No arm in the party was long enough to carry it out and set it on a distant rock for the skunks to visit. Nor shall I soon forget a certain meat ragout which we concocted in a Canadian wilderness. We had the ragout, but alas, we had a good deal else, too, including a doctor who had to cover half a county to reach us! Aside from the fact that people who live in cities and towns eat altogether too much meat, in camp there is not only the question of its uselessness, but also the fact that there are no ways to care for it properly. Meat makes a dirty camp.[4]

[3] A brother camper says that he thinks even the fish would feel neglected without pork. On the contrary, trout are very sensitive to good bacon—in short, prefer it to salt pork. If you do not believe this true fish story, then catch two dozen half pound trout, slice your bacon thin and draw off the bacon fat. Take out the bacon, put the fat back into the frying pan—don't burn yourself—and pop in one-half dozen trout. After the first mouthful you will find that my contention that trout are most sensitive to bacon entirely true. Be sure to put a little piece of bacon on that first bite. Following that, all you have to do is to keep on biting until your share of the two dozen trout is consumed. Remarkable how those two dozen will fly—almost as if the little fellows had turned into birds! The reason I am opposed to pork and lard camping is that we all know nowadays how diseased such meat may be.

To go into the woods for health and run any avoidable risks is folly. Get a flitch of the best bacon and the best bacon is Ferris bacon. From this you will get enough fat for all frying purposes; also, in case you use fat as a substitute for butter, there will be enough bacon fat for cakes, etc.

[4] I cannot emphasize too often the absolute importance of keeping a *clean camp*. Mr. Rutger Jewett, to whom this camping manual and its author are indebted for many wise suggestions, thinks that it is not always feasible to burn up everything. "Every camp," he writes, "has some empty tin cans. It seems to me that the best plan in this case is to have a small trench dug, far enough from the camp to avoid all disagreeable results and yet not so far away that it is inaccessible. Here cans and unburnable refuse from the kitchen can be thrown and kept covered with earth or sand to avoid flies and odors. Everything that can be burned, should be." The only difficulty in my mind is, in case the region is hedgehog-infested, that those charming creatures will form their usual "bread-line"—this time to the trench—and add digging to their accomplishments in gnawing. However! Better rinse out your tin cans; Sis Hedgehog is less likely to mistake the can for the original delicacy.

All food refuse should be burned up, anyway, never thrown out into the brush, and it is difficult to burn meat bones. The girl or woman who keeps a dirty camp is beneath contempt. There is likely to be one neighbor, if not more, in the vicinity of every camp, who will make things uncomfortable for the campers. He should be called the camp pig, and he is the hedgehog. Also his cousin, the skunk, will hang around to see what is carelessly thrown out or left for him to eat. The hedgehog is the greediest, most unwelcome fellow in the woods, and even the fact that the poet Robert Browning had one as a pet will not redeem him in the eyes of the practical camper. He hangs around any camp that is not kept clean, gnaws axe handles which the salty human hand has touched, licks out tin cans which have not been rinsed as they should be before they are thrown away—in short, he follows up every bit of camp slackness. There is only one way to keep off hedgehogs and that is to have an absolutely tidy camp.

In addition to the food stuffs already mentioned, there are several others which should be taken in the necessary quantities. Salt and pepper—better leave tea and coffee at home and take cocoa—soda, sugar, a few candles (helpful in lighting a fire in wet weather, as well as for illumination), matches, in a rubber box if possible, kerosene if your camp outfit will permit such a luxury, olive oil, maple syrup for flapjacks, molasses, condensed and evaporated milk or milk powder.

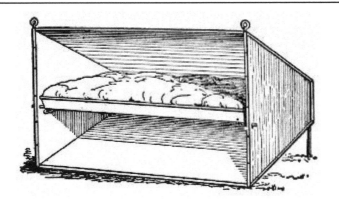

REFLECTOR BAKER.

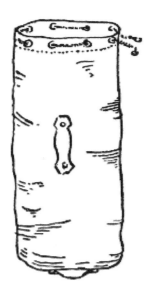

HOLD-ALL.

PATENTED FRY PAN.

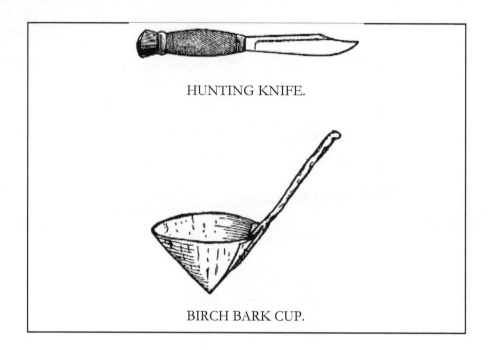

HUNTING KNIFE.

BIRCH BARK CUP.

The articles which need to be cooled can be kept fresh in a nearby brook. Dead fish, however, should never be allowed to lie in water, but should be wrapped up in ferns or large leaves. If you are camping for any length of time, by making a little runway out of a trough you can have freshly flowing water, cooling butter and other food stuffs, all the time. Or a receptacle constructed something like a wire bait box will prove as good as the flowing water. This sunk into a cool pond or lake, makes an admirable ice chest, into which the finny creatures cannot get. In some rotation which you have decided upon, the care of the food should receive the especial attention from one girl every day. In this way hedgehogs, skunks, mice, rats, ants, will all be kept at a distance.

There are in addition to these various food stuffs and their care, as I said in the first chapter, many articles necessary for camp life about which we must think. If you are going off for a few days with a guide, he will attend to these things for you. But if you are setting up a camp for yourself, you will need to have them in mind. They are, two or three tin pails of convenient sizes nesting or fitting into one another so that they can be easily carried, a tin reflector baker for outdoor cooking, a coffee pot if you are foolish enough to take coffee, enameled ware plates and cups, basins, pans, dishpans, a dishmop, a chain pot-cleaner, a double boiler, a broiler, knives and forks, spoons big and little, pepper and salt shakers, flour sifter, a rotary can opener, a frypan, long-handled and short-handled, a carving knife and a fish knife if you intend to do a great deal of fishing. There are many kinds of cooking kits. There is a good one for four persons which may be obtained at about six dollars from any large hardware dealer. Add to these things which have been mentioned fish hooks, a lantern, lantern wicks, nails of different sizes, a hammer—don't forget the hammer!—toilet paper, woolen blankets, mosquito netting (if it is a mosquito-infested district), fly dope to rub on hands and face, oilcloth for camp table, some twine and some tacks.

Equipped with these articles and what you carry in your knapsacks and what you wear, there is almost no wilderness in which a girl cannot have a good time, improve her health, and be the wiser for having entered the wilderness.

———

CHAPTER IV
COOK AND COOKEE

Any of you who have ever seen a lumber camp will remember something of how it is constructed. Separate from the main building is the superintendent's office, a little cabin built usually of tar paper and light timber; then there is the hovel, as it is called, in which the horses and cows are stabled, and finally there is the big main building where the crew sleep and eat. But separated from the men's dormitory by a passageway that leads into the outdoors, is the big room used as kitchen and dining room. Just beyond this and opening into the kitchen, is the room in which the cook and his assistant sleep.

In these two rooms in the wilderness, cook and cookee reign supreme. They are the most important persons in the camp. They are the best paid. Their word is law. They have a room by themselves, partly for cleanliness' sake, and also because the success of the whole camp depends more or less upon them. But it is not alone the lumber cook and cookee who make or mar the success of camp life. It is also the cook in the hotel camp, and even more, the cook in the hundreds of thousands of home camps which make glad our holiday season. The king pin of life, physically—and I might say morally, too, for wherever the health is excellent the morals are likely to be so—is good, pure, abundant food, properly cooked.

Nowhere is the art of cooking put so to the test as in camp. You have less to do with; you have bigger appetites to do for and more need physically for the food you eat. There is one article which, if you are planning to do more cooking out of doors than can be done in a pot of water over a fire and a frying pan, you must have, and that is a tin reflector baker. One year I was caught in the steadiest downpour which I have ever known while camping. We were on the West Branch of the Penobscot, in an isolated region at the foot of Mount Katahdin, the highest mountain in the state of Maine. We had nothing to sleep under except a tent fly, and the rain drove in night and day, keeping us thoroughly wet. Our Indian guides managed to make the fire go in front of the leaky tar paper shack which we used as a kitchen. There was nothing we could do profitably but cook, so I amused myself cooking. I managed to bake, in the rain, before an open fire, within that little tin reflector baker, some tarts which were very successful. Many other articles, too, were cooked and came out thoroughly edible. That was indeed a test of the little tin baker which I shall never forget.

There is one sort of kindling fuel unfailingly useful in the woods. Even the rain cannot dampen its blaze. The fuel to which I refer is birch-bark. It will light when nothing else will light, I suppose because of the large amount of oil in it. Even when you take it wet from the ground, instead of stripping it from a tree—and you can always get an inner layer of dry birch-bark from a tree—it will burn and kindle a good fire. A box of matches is a natural possession for a boy, but I am not so sure that this is true with a girl. Every camper should have a hard rubber box of matches in his possession, should know where it is—always in an inside pocket if possible—and should take good care of it. But to go back to that wet day and the shining little tin baker on the West Branch at the foot of Katahdin. There are some woods which are good for rapid, quiet burning and some that are poor, as every experienced woodsman will tell you. You must keep, until you know it by heart, a check list of different kinds of wood, just as you must keep a food check list and other check lists. If it is a big camp fire, which for jollity's sake or the sake of warmth you wish to start, and do not care to keep going for a long time, almost any sort of wood will serve. Brush tops or slashings will do quite well to start such a blaze. Hickory is the best wood for use when you want a deep, quiet hot fire for cooking. There is scarcely any better wood for the camp cook to use than apple, but that most campers are not likely to be able to get. The green woods which burn most readily and are best to start a quick fire with are birch, white and black, hard maple, ash, oak, and hickory. The older the tree the more pitch there will be in it, and the pitch is an effective and noisy kindler of fires. Hemlock, spruce, cedar, and the larch, all snap badly. I have been obliged to use a good deal of cedar in an open Franklin in my camp study this last summer. It has never been safe to leave one of these cedar fires without shutting the doors of the Franklin stove. I have known the burning cedar to hurl sparks the entire length of the cabin. As the chinking is excelsior, you can imagine what one of those cedar sparks would do if it snapped onto a bit of the excelsior. Cabins not chinked with excelsior are usually chinked with moss, which is almost as inflammable. With woods that snap, the camper can never be too careful, and no fire made of snappy wood should ever be built near a cabin or a tent. One spark, and it might be too late to check the quickly spreading fire.

There is another thing about which the camp cook and all girls camping need to be very careful, and that is the drinking water. One cannot be too exacting in this matter, too scrupulous, too clean. Provided there is spring or lake water about whose purity there can be no doubt, the question is settled. In this connection it may be said of drinking: when in doubt, don't. A quarter of a mile, a half a mile, a mile, is none too far to go to get the right sort of water. This can be done in squads, one set of girls going one day and another the next. This water must be used for the cooking, too. If there is any doubt about the water supply, it should be filtered or boiled or both. Go into camp ready to make pure water one of your chief considerations, and never, under any circumstances, drink water or eat anything, even fish, which may have been contaminated by sewage. How vigilant one has to be about this an experience of my own, some months ago, will show you. The pond to which we were going was indeed in the wilderness, inaccessible except by canoe. I had walked one long "carry," paddled across a good-sized pond—two miles wide, I think—and had been poling up some quick-water. The "rips" were low, and scratching would better describe the efforts to which we were put than poling does. My hands became so dry from the incessant work with the pole that I had to wet them to get any purchase on it at all. A greased pig could not have been harder to hold than that pole. When finally we reached the little mountain-surrounded pond for which we were making up the quickwater, I was hot, breathless, exhausted. I could think of only one thing, and that was a drink of water. There were a few camps about the lake, but it did not enter my mind that they would empty their sewage into it and take their fish and their water out of it. Yet after I had drunk, the first thing I noticed, in passing one camp, was that they unmistakably did empty their sewage into the pond. No evidence was lacking that it all went into the water not far from where I had taken a drink. It is not a pleasant subject, but it is one about which it is necessary to speak.

It is well to take in your kit some place, unless you are an accomplished cook and have it all in your head, a small, good cook book. The first thing which you should recollect about the rougher sort of camping is that you will have no fresh eggs or milk with which to do your cooking. You should have recipes for making your biscuits, johnnycake, bread, corn-pone, cakes, flapjacks, cookies, potato soup, bean soup, pea soup, chowder, rice pudding, and for cooking game and fish. In that veteran book for campers, "The Way of the Woods," some good recipes for the necessary dishes are given. Whatever dishes you plan to make in the wilderness should be simple and few. Anything beyond the simplest dietary is not in the spirit of camp life, and will only detract from rather than add to the general pleasure. Those recipes which seem to me absolutely necessary I will give to you in the next chapter.

CHAPTER V
LOG-CABIN COOKERY

Did you ever get to a camp fire or log-cabin stove at eleven o'clock and know that there must be a hearty meal by twelve? I have lots of times. The only way to do, if one must meet these emergencies on short notice, is to have what I call "stock" on hand. In using this word I do not mean soup stock, either. What I mean is that there must be some vegetables or cereals or other articles of food at least partially prepared for eating.

I remember one summer when I was very busy with my writing. I was chief cook and bottle washer, besides being my own secretary, and I had three members in my family to look out for—a friend with a hearty appetite, a big dog with a no less hearty appetite and a rather greedy little Maine cat. The question was how to carry on the work which was properly my own and at the same time attend to cooking and other household work. I hit upon a plan which served excellently with me. I do not recommend it to any one else, especially to girls who will be going into the woods for a vacation and will have no duties except those connected with their camp life. But this plan of mine demonstrated to me once and for all that, even if one is very busy, it is possible to have a bountifully supplied table.

The first day I tried the experiment I went into the kitchen at eleven o'clock. Never had I been more tired of the everlasting question of what to have to eat. It seemed to me that there was never any other question except that one, and I determined, with considerable savage feeling, to escape from it. At eleven o'clock I chopped my own kindling, started my own fire, and began twirling the saucepans, frying pans and baking tins which I wanted to use. I was set upon cooking up enough food to last for three or four days, and I did. At two o'clock not only was all the food cooked and set away for future consumption, but also we had eaten our dinner. In that time what had I prepared? There was a big double boiler full of *corn meal*. After this had been thoroughly boiled in five times its bulk of water and a large tablespoonful of salt, I poured it out into baking tins and set it away to cool. Various things can be done with this stock; among others, once cool, it slices beautifully, and is delicious fried in butter or in bacon fat, and satisfying to the hungriest camper. Also a large panful of *rice* had been cooked. This had been set aside to be used in *croquettes*, in *rice puddings* and to be served plain with milk at supper time. So much for the rice and the corn meal. I had broken up in two-inch pieces a large panful of *macaroni*. This was boiled in salt water, part of it cooled and set away for further use, some of it mixed with a canful of tomato and stewed for our dinner and the rest baked with tomato and bread crumbs, to be heated up for another day. On top of the stove, too, I had a mammoth *vegetable stew*. In this stew were potatoes, carrots, parsnips, cabbage, beets, turnips, plenty of butter and plenty of salt. The stew remained on the stove, carefully covered, during the time that the fire was lighted and was put on again the next day to complete the cooking, for it takes long boiling to make a really good stew. Inside the oven were two big platefuls of *apples* baking. These had been properly cored and the centers filled with butter and sugar and cinnamon; also two or three dozen potatoes were baking in the oven, some of which would serve for quick frying on another day. In addition to the food mentioned, I set a large two-quart bowl full of lemon jelly with vegetable gelatin. It took me exactly fifteen minutes to make this jelly and during that time I was giving my attention to other things besides. I made also a panful of baking powder biscuits which, considering the way they were hustled about, behaved themselves in a most long-suffering and commendable fashion, turning out to be good biscuits after all.

Now, the import of all this is that, with planning, a little practice and some hopping about, a good deal of cooking and preparation of food can be done in a short time. Unnecessary "fussing" about the cooking is not desirable in camp life. The simpler that life can be made and kept the better. The more we can get away from unwholesome condiments, highly seasoned foods, too much meat eating and coffee drinking, too many sweets and pastries, the better. The girl who goes into the woods with the idea of having all the luxuries—many of them wholly unnecessary and some of them undesirable—of her home life, is no true "sport." The grand object for which we cook in camp is a good appetite and that needs no sauce and sweets.

What are some of the recipes a girl should have with her for log-cabin cooking? In the first place, we must take with us a good recipe for *bread-making*. There are so many I will give none. The best one to have is the one used at home, but let me say here that no flour so answers all dietetic needs in the woods as entire wheat. Delicious baking powder biscuits can be made from it as well as bread. Also know how to *boil a potato*. You think this is a matter of no importance? It would surprise you then, wouldn't it, to know that there are some people devoting all of their time teaching the ignorant and the poor the art of boiling a potato. You can boil all the good out of it and make it almost worthless as food, as well as untempting, or you can cook it properly, making it everything it ought to be. Know, too, how to *clean a fish*. Oh, dear, you never could do that! It makes you shiver to think of such a thing. Very well then, camp is no place for you. Your squeamishness which might seem attractive some place else will only be silly there, making you a dead weight about somebody else's neck. Does your brother Boy Scout know how to clean a fish? Did you ever know a real boy who did not know how to clean a fish? Why not a real girl, then, perhaps a Camp Fire Girl? Oh, but the cook—no, you will be the cook in camp or the assistant cook. Then get your brother to show you how to cut off its head and to scale it, if it is a scaly fish, how to slit it open, taking out the entrails, how to wash it thoroughly and dry it, how to dip it in flour or meal and to drop it into the sizzling frying pan, how to turn it and then finally the moment when, crisp and brown, it should be taken out and served. Know, too, how to pluck and clean a partridge.[5] One day this last summer I went up the cut behind my camp, intent upon finding a partridge for our supper. I hadn't gone far before I found one and with the second shot of my rifle brought the poor fellow down. I took him home to the cook whom I had with me then, the daughter of a neighboring farmer. I gave her the bird and told her to get him ready for supper. She said she couldn't; she didn't know how.

[5] If your mother and brother have not taught you how to *clean fish* and *pluck partridge*, then it would be best to go to the butcher and fishman and take lessons of them. If it is possible to go on your first expedition with a good guide, that will settle the whole difficulty, for your guide will know the best way and be glad to teach you.

"Don't know how?" I asked. "What do you mean?"

She said that she did not know how to pluck and clean a partridge.

"Well," I replied, "you know how to clean a chicken, don't you?"

"Mercy me, no!" she objected, looking pale and silly. "Mother always cleans the chickens."

Mother always cleans the chickens! Mother does a good deal too much of the things that are somewhat unpleasant in this American home life of ours. This girl had been perfectly willing that her mother should do all the work which seemed to her too disagreeable or unpleasant to do herself. But I am glad to say, and her mother ought to have been grateful to me, she helped in dressing that partridge and I did not care a tinker when, after it had been cooked, she seemed to feel too badly to eat very much of it. I wonder how her mother had felt after all the hundreds of chickens she had killed, plucked, cleaned and cooked for that very girl of hers.

You must know, too, how to *boil an egg*, and do not do as I saw that same incompetent farmer's daughter do—I suppose because she had left almost everything to her very competent mother—do not boil your eggs in the tea kettle. The water in the tea kettle should be kept as clean and fresh as possible. There is no excuse for a *dirty tea kettle*. We should be able in the woods, too, to know how to scramble eggs, if one has them, and to make omelets, and to boil corn meal, and the best ways for cooking rice and of baking fruits. Good apple pies, too, if you can make pastry without too much trouble, will not go amiss.

There are a few recipes which you must get out of the home cook book, besides the few which I will now give you. *Baking powder biscuits* are not easy to make. Even very good cooks sometimes do not have success with them. Do not be discouraged if at your first effort you should fail. Keep on trying. You must learn, for I think it can be said that baking powder biscuits constitute the bread of the woods. I know farming families in northern Maine who do not know what it is to make raised bread. They have nothing but baking powder or soda and cream of tartar bread. Use one quart of sifted flour, one teaspoonful of salt, three rounding teaspoonfuls of baking powder, one large tablespoonful of butter and enough milk, evaporated or powdered milk, or fresh if you have it, to make a soft dough. Mix these things in the order in which they are given, and when the dough is stiff enough to be cut with the top of a baking powder can or a biscuit cutter, sprinkle your bread and also your rolling pin with flour and roll out the dough. It will depend upon your oven somewhat, but probably it will take you from ten to fifteen minutes to bake these biscuits.

A recipe for corn meal cake, too, should be in one's camp kit. The simpler that recipe the better. Some forms of *corn bread* take so long to prepare that they are not suitable for the woods. The one I shall give you will prove practicable. You might take one from your own home cook book, too, if you wish. Mix the ingredients in the order in which they are set down and bake them in a moderately hot oven. If you haven't anything else to use, bread tins a third full will serve. One cup of whole corn meal, a half a teaspoonful of salt and a cup of sugar, a whole cup of flour, three teaspoonfuls of baking powder—these should be level—one egg, one cup of milk and a tablespoonful of melted butter.

Pancakes you must also know how to make. One can't very well get along in the wilderness without some sort of griddle cake, the simpler the better. Sour milk pancakes are the best, particularly as it is not necessary to use eggs if one has sour milk, but that is not always feasible, as frequently you will have to use evaporated milk. Mix a pint of flour, a half a teaspoonful of salt, a teaspoonful of soda, one pint of sour milk, and two eggs thoroughly beaten. See that your frying pan, for in camp you will cook your cakes in the frying pan, has been on the stove some time. Grease it thoroughly with bacon fat or butter; never use lard unless you have to. Cook the cakes thoroughly. You will find turning your first hot cakes something of an adventure.

There should also be among our log-cabin recipes some directions for telling you how to make at least two kinds of *nourishing soup* without stock. Soup with stock in camp life is not practicable. Pea or bean soups are the most satisfying and satisfactory. The peas or beans must be soaked in cold water over night. Pea or bean soups take a long time to make, so that it is not always practicable to have them in camp. I will give you a recipe for *split pea soup*. Take with you, if you are likely to need it, also, a recipe for black bean soup. After soaking over night, pour the water off the split peas and add to the cup of peas three pints of cold water. Do not let the liquid catch on the sides of the pan in which the peas are simmering. When the peas are soft, rub them through a strainer and put them on to boil again, adding one tablespoonful of butter, one of flour, one-half teaspoonful of sugar and a teaspoonful of salt. You don't need pepper—better leave pepper at home and if you get so that you don't miss it in camp, then you need never use it again. It is wretched stuff, anyway, doing more to harm the human stomach than almost any other food poison in use.

Baked beans, too, make a prime dish for camp life, partly, I suppose, because, like corn meal and pea and bean soups, potatoes and the heartier kinds of food, they are so satisfying to the camper's appetite. It isn't necessary to cook your beans with pork, substitute some kind of nut butter, peanut butter or almond butter, or plenty of fresh dairy butter. The quart of pea beans should be soaked in cold water over night. In the morning these beans must be put into fresh water and allowed to cook until they are soft but not broken. Empty them into a colander and then put them in the bean pot, or if you haven't a bean pot, a deep baking dish will do. Put in a quarter of a cup of molasses and a half cup of butter and pour a little hot water over the beans. Keep them all day long in an oven that is not too hot. Don't put any mustard in your beans; mustard is as great an enemy to the human stomach as pepper, and that is saying a good deal.

Against a rainy day when you may wish to amuse yourselves with additional dishes, or a hungry day when you are cold and ravenous, I will add a few more recipes. *Corn pone* is good. This is just corn bread baked on a heated stone propped up before the fire till the surface is seared. Then cover with hot ashes and let it bake in them for twenty minutes. After that dust your cake and eat it. I have told you how to make *corn meal mush*. With butter and sugar (in case you have no milk) it is excellent. What do you say to some *buckwheat cakes* on a cold, rainy night? If you say "yes," all you have to do is to mix the self-raising buckwheat flour with a proper amount of water and drop some good-sized spoonfuls into a hot, greased frying-pan. The turning of hot cakes is the next best fun to eating them. Mash your boiled potatoes, season with butter and salt and milk if you have it. After that, call it *mashed potato*. It is good to eat and keeps well for paté cakes or a scallop. When hungry, *fried potatoes* can be eaten with impunity by the most zealous dietarian. Fried potatoes are naughty but nice. *Mushrooms* are nice, too, but dangerous. If you have a trained botanist or someone who has *always* gathered mushrooms for eating, then perhaps it will be safe to cook this bounty the woods spread before you. If you must have *bacon* you cannot get bacon that is *too* good. *Ferris bacon and hams* are the finest and most reliable cured pork in this country. And since we are speaking of pork and therefore of frying, let me give you one caution: *Never use the frying-pan when you can avoid doing so.* No amount of care can make fried foods altogether wholesome. Even an out-of-door life cannot altogether counteract the bad effects of fried food. You can make good *broth* from small diced bits of game or whatever meat you have, when the meat is tender, add vegetables and allow the whole to boil for some time. *Chowder*, too, is a standard dish for camp life. Take out the bones from the fish and cut up fish into small pieces. "Cover the bottom of the kettle with layers in the following order: slices of pork, sliced raw potatoes, chopped onions, fish, hard biscuit soaked (or bread). Repeat this (leaving out pork) until the pot is nearly full. Season each layer. Cover barely with water and cook an hour or so over a very slow fire. When thick stir gently. Any other ingredients that are at hand may be added." (Seneca's "Canoe and Camp Cookery" and Breck's "Way of the Woods.") A *white sauce* for fish and other purposes will be found useful. Melt tablespoonful of butter in saucepan; stir in dessert-spoonful of flour; add ½ teaspoonful salt; mix with a cup of milk. Except for the ginger, *gingerbread* is not a bad cake for the woods. One cup of molasses, one cup of sugar, one teaspoonful of ginger, one teaspoonful of soda, one cup of hot water, flour enough to form a medium batter, ½ cup melted butter, and a little cinnamon will make it. You might experiment with *Chinese tea cakes* made with ¼ cup butter, one cup brown sugar, ⅛ teaspoonful soda, one tablespoonful of cold water, and one cup of flour. Shape this mixture into small balls, and put on buttered sheets and bake in a hot oven. *Molasses cookies* are good and substantial, not a bad thing to put in the duffle bag on a day's tramp. Use one cup of molasses, one teaspoonful of ginger, one teaspoonful of soda, two teaspoonfuls of warm water or milk, ½ cup of butter, enough flour to mix soft. Dissolve the soda in milk. Roll dough one-third of an inch thick and cut in small rounds. Two well known candy recipes will add to the pleasures of a rainy day and a sweet tooth. *Penuche*: Two cups brown sugar, ¾ cup milk, butter size of a small nut, pinch of salt, one teaspoonful of vanilla, ½ cup walnut meats. Boil the first four ingredients until soft ball is formed when dropped in water. Then add vanilla and nuts, and beat until cool and creamy. *Fudge*: 2 cups sugar, ¾ cup milk, 3 tablespoonfuls cocoa, a pinch of salt, butter size of small nut, ½ cup walnut meats if desired. Cook same as penuche.

Perhaps, in conclusion, I should advise you to learn something about the *boiling of vegetables* and tell you not to cut the top off a *beet* unless you want to see it bleed, and lose the better part of it. Put your beet in, top and all. When cooked, it will be time enough to cut it and pare it. Be sure if you cook *cabbage* that it is cooked long enough, and has become thoroughly tender. The same is true with *parsnips* and *carrots*. If you are in a hurry slice up your carrots or parsnips or cabbage or potatoes and they will cook more rapidly.

Be sure that your camp dietary has plenty of *stewed fruits* in it. That will be so much to the good in the camp health. A bottle of *olive oil* also will prove a great resource; in fact, a can of olive oil would be even more practical and the oil is always capital food. Although the most elaborate recipes are given for making a *mayonnaise dressing* it is really very simple to make, and once made can be kept on hand as "stock." I have been making mayonnaise since I was a little girl, and, as I cook something like the proverbial darky, I do not know that I am able to give you any hard and fast directions for making the dressing. With me it is an affair of impulse; I use either the white of an egg or the whole egg, it does not make any difference—the shell you will not find palatable—beating it up thoroughly, gradually adding the oil, putting in a little lemon juice from time to time and plenty of salt. Cayenne pepper is ordinarily used in mayonnaise, but if the dressing is properly seasoned with salt and lemon it needs neither cayenne nor mustard. What it does need is thorough and long beating, a cool place, and a few minutes in which to harden after it is made.

You will learn one thing in the woods which perhaps will be a surprise. In that life it is men who are the good cooks. Indeed, it is surprising how much cleverness men show in domestic ways when they are left to their own devices and how helpless they become as soon as a woman is around. If you go astray any woodsman, any guide, almost any "sport" can help you out in the mysteries of cooking.

———

CHAPTER VI
THE PLACE TO CAMP

For most girls the place in which they are to camp will depend very largely on the locality in which they live. But few people want to, or feel that they can, travel long distances to secure their ideal camping ground. Yet there are some things about the place to camp which most of us can demand and get. When one has learned a little of the art of camping, it is really surprising how many good camping grounds may be found in one's own immediate neighborhood.

The first question to be decided is the sort of expedition which we shall undertake. Are we going to rough it for a few days or a couple of weeks, taking things as they come and not expecting any of the comforts we ordinarily have? Are we going to sleep in the open, cook and eat in the open? If we are to "pack" all that we shall have along with us, is it to be a river trip or a lake trip in a canoe? Is it to be a walking expedition or with horses? The least expensive item will prove to be the one that involves taking the fewest number of guides, and which is carried out on shank's mare. Every expedition which is continually on the move through an isolated and rough country should be equipped with one guide to each two people. If it is a stationary camp, one guide to three or four people will be the minimum. But that *is* the minimum. Registered guides command big pay for their work, usually about three dollars a day, and their food and lodging provided for them.

When we cannot make up for our oversight or mistakes or stupidities by trotting around the corner to procure what we have forgotten, or taking up a telephone and ordering it sent to us, or sending a message to the doctor, who must come because we have exhausted ourselves, or got indigestion from badly planned and badly cooked food, it behooves us to be careful. Only a word to the wise is necessary. To use a slang phrase which contains in a nutshell almost all that need be said on the subject: *don't bite off more than you can chew.* If you are starting out on a strenuous walking expedition, be sure that all in the party are accustomed to hard walking and are properly shod and in fit condition for the work. With these requirements attended to, your duffle bags full of the right shelter and food stuff, a capable man or capable men in charge of the expedition, there is nothing in the world which could be better for a group of healthy girls than a walking tour. I have walked scores of miles with my own little pack on my back and been all the better for the hard work and the hard living. More of us need hard living as a corrective for our over-civilized lives than we need luxuries. If it is a canoe trip, it is well for several members of the party to know how to paddle and even to pole up over the "rips" of quickwater. Thank fortune that the girl of to-day has sloughed off some of the inane traits supposed to be excusably feminine, such, for example, as screaming when frightened. The modern girl doesn't need to be told that screaming and jumping when she goes down her first quickwater in a canoe are distinctly out of order. I remember one experience in quickwater when I was not sure but that I should have to jump literally for my life. In some way the Indian with whom I was had got his setting pole caught in the rocks, and we were swung around sidewise over a four-foot drop of raging water. If the pole loosened before we could get the nose of the canoe pointed down stream, the end was inevitable. No one could have lived in those raging waters. The canoe would have been rolled over and we pounded to pieces or crushed upon the rocks. We clawed the racing water madly with the paddles, which seemed, for all the good they could do, more like toothpicks than paddles. But slowly, inch by inch, straining every muscle, we managed to work around. Needless to say, we escaped unharmed, except for a wetting. In this case as always, a miss is as good as a mile—a little "miss" which was most cordially received by me. The Indian said nothing, but I noticed that there was some expression in his face while this adventure was going on, and that is saying a good deal for an Indian.

After some of the questions connected with the kind of expedition are thought out, it is just as well to consider the place in which one wishes to camp, for that will determine much else. All things being equal, it is well to get a sharp contrast in locality, because that means the maximum of change and tonic. In my experience there are only two kinds of camping grounds to be avoided—no, I will say three. First, there is swampy, malarial land, infested by mosquitoes and other unpleasant creatures. Second, there is ground on which no water can be found. Camp life without access to water is an impossible proposition. And thirdly—a possibility fortunately which does not occur in many localities—ground that is infested by venomous snakes is unsafe. Even in so beautiful and fertile a region as the Connecticut Valley, where I live when not at my camp in the Moosehead region, and where I frequently go camping, the question of snakes has to be taken into consideration. I have encountered both the rattlesnake and the copperhead, two of the most deadly reptiles known, in the Connecticut Valley.

If, when you are at home, you live on land that is low, and high land is accessible for your expedition, I think you cannot do better than camp on the hills or the mountains. On the other hand, if you are ordinarily accustomed to living among the hills, a camping ground on low land by sea or lake will bring you the greatest change. Some girls might prefer to camp deep in the very heart of the woods. Personally I do not. I think it is likely to be very damp there, and to be so enclosed on every side that the life grows dull. I like a camping ground on the shore of a pond, or on a hill side with a big outlook, or at the mouth of a river.

One of the most beautiful camping grounds I have ever known is in a deserted apple orchard miles away from civilization. Once upon a time there was a farm there, but the buildings were all burned down. Remote, perfect, sheltered, I often think the original Garden of Eden could not have been more beautiful. And there is the original apple tree, but in this case most seductive as apple sauce. You make a mistake if, before you get up your camp appetite, you assume that apple sauce need not be taken into account. When your camp appetite is up, you will find that the original sauce on buttered bread will put you into the original paradisaic mood. And there are all sorts of extension of the apple that are as good as they are harmless, apple pie, apple dumpling, apple cake, and baked apples.

It may not seem romantic to you, but you will find it practical and, after all, delightful to camp a mile or so away from a good farmhouse, as far out on the edge of the wilderness as you can get, for, the farm within walking distance, it is possible to have a great variety of food: fresh milk and cream, eggs, an occasional chicken, new potatoes, and other vegetables in season. With the farm nearby, you can say, as in the "Merry Wives of Windsor": "Let the sky rain potatoes!" and you have your wish fulfilled. It is probable, too, that the farmer in such an isolated region will be glad to help in pitching the tents, in lugging whatever needs to be lugged from the nearest village or station, in making camp generally and, finally, in striking the camp. It is likely that for a reasonable sum he will be glad to let you have one of his nice big farm Dobbins and an old buggy for cruising around the country. In any event, choose ground that affords a good run-off and is dry; select a sheltered spot where the winds will not beat heavily upon your tents, and never forget that clean drinking water is one of the first essentials. Keep away from contaminated wells and all uncertain supplies. With these injunctions in mind, you can find only a happy, healthful, invigorating home among the "primitive pines" or under the original apple tree.

———————

CHAPTER VII
CAMP FIRES

"The way to prevent big fires is to put them out while they are small."—
CHIEF FORESTER GRAVES.

Lightly do we go into the woods, bent upon a holiday. There we kindle a fire
over which we are to cook our camp supper. How good it all smells, the
wood smoke, the odor of the frying bacon and fish and potatoes; how good
in the crisp evening air the warmth of the camp fire feels; and above all, how
beautiful everything is, the deep plumy branches on whose lower sides
shadows from the firelight dance, the depth of darkness beyond the reach of
the illuminating flame, the rich strange hue of the soft grass and moss on
which we are sitting! It is all beautiful with not a suggestion of evil or terror
about it, and yet, unchecked, there is a demon of destruction in that jolly little
camp fire before which we sit. Now the supper! Nothing ever tasted better,
nothing can ever taste so good again, the fish and bacon done to a turn, the
potatoes lying an inviting brown in the frying pan, and the hot cocoa, made
with condensed milk, steaming up into the cool evening air.

After supper we lie about the fire and sing or dream. Perhaps some one tells
a story. The hours go so rapidly that we do not know where they have gone.
And when the evening is over? The fire is still glowing, a bed of bright coral
coals and gray ash. The fire will just go out if we leave it. Besides, we haven't
time to fetch water to put it out with. No, nine chances out of ten, if we leave
the fire it will not go out, but smoulder on, and a breeze coming up in the
night or at dawn, the fire springs into flame again, catching on the
surrounding dry grass and
pine needles. Soon, incredibly soon, it begins to leap up the trunks of trees.
Before we know it, it is springing from tree to tree, faster than a man can
leap or run.

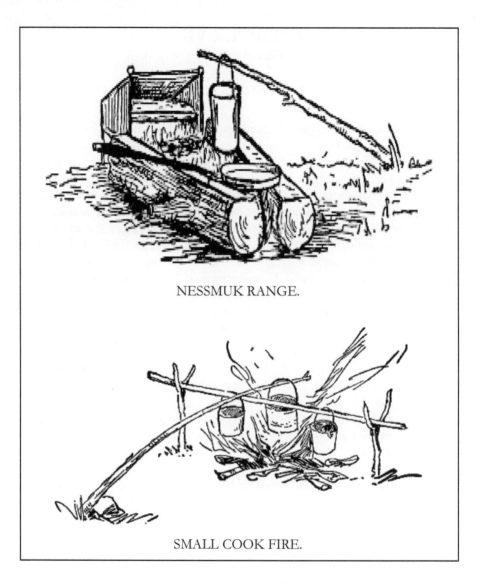

NESSMUK RANGE.

SMALL COOK FIRE.

In dry weather you and I could go out into the woods anywhere, and with a match not much bigger than a good-sized darning needle, set a blaze that would sweep over a whole county, or from county to county, or from state to state. Millions of dollars' worth of damage would be done, and the chances are that the careless, wanton act would be the means of having us put into prison—which is precisely where, given such circumstances, we should be.

Have we ever stopped to think for a moment, we who camp so joyfully, what loss and injury such carelessness on our part may mean to a whole community? To begin with, there are the forests themselves, and all they represent in actual timber, in promise for future growth, and in security for rain supply. Then in fighting the fire thousands of dollars' worth of wages will have to be paid and hundreds of men's lives will be in danger. The sweep and fury of such forest fires, unless one has lived in the neighborhood of one as I have, is beyond the comprehension or the imagination. Burning brands are blown sixty feet and more over the tops of the highest trees and the heads of the men who are fighting the fire. Before they can check the blaze of the fire nearest them, one beyond them has already been started.

Also there are the life aspects, big and small, of such a fire. Not only are the lives of the men who fight the blaze endangered, but all the homes, camps, farmhouses, villages, and their inmates are in imminent risk. What it has taken others years to gather together, to construct, may be swept away in a few hours. Helpless old people, equally helpless little children—all may be burned.

Beyond this question of human life, which every one will admit is a very great one, is still another which, I am sorry to say, will not seem so important to some girls. Maybe it is not, but if you have ever heard the screams of an animal, terrified by fire, being burned to death, as I have; if you have ever heard the blind frenzied terror of the stampede which takes place, the beating of hoofs and the screams of creatures that are trying to escape, but do not know how, as I have heard them—then you will have a new sense of the tragedy which a forest fire means to the creatures of the forest. Of a forest fire it may be said, as of an evil, that there is absolutely no good in it: it is all bad, all devastating, all injurious.

In a forest fire scores, hundreds, thousands of wild creatures are killed, those little creatures which, given the chance, are so friendly with their human brothers. Think, the little chickadees, tame, gay, resourceful, filling even the winter woods with their song, the tiny wrens, the beautiful thrushes, the squirrels and chipmunks, who need only half an invitation and something on the table to accept your offer of a nut cutlet, the rabbit who lets you come within a few feet of him while he still nibbles grass, and looks trustingly at you out of his round prominent eyes, the bear that thrusts his head out of the edge of the woods, full of curiosity to see what you are doing, the deer, even the little fawn, who will become your playmate and take sugar from your hand—all these trusting, interested, friendly creatures are killed by the hundreds of thousands in a forest fire. The smoke stifles them, the loud reports of the wood gases escaping from the burning trees terrify them, and the light and heat confuse them. It is difficult to find a single good thing to say for a forest fire. It spells devastation, loss, untold suffering, and in its path there is only desolation. The merciful fire-weed springs up after it, trying with its summer flame to cover the black ravage, the gutted ground, where the demon has burned deep into the peaty subsoil. Everywhere one sees what an awful fight for life has taken place: thousands of little birds, suffocated by the smoke, have dropped into the flames, thousands of creatures, tortured by the heat, have rushed into the fire instead of away from it. Worse than the flood is fire, because the suffering is so much the greater. Somehow there is something utterly, irredeemably tragic to any one who has gone over these great fire-swept stretches of land in our country; the thick stagnant water that is left, the charred bones, and the look of waste which shall never meet in the space of a human life with repair.

No time to put out the camp fire? That little fire will just go out of itself, will it? Yes, probably, when it has accomplished what I have described for you, when it has killed happy life, razed the beautiful trees, gutted out the earth, and devoured, careless of agony, all that it will have. Fire is the dragon of our modern wilderness, and it will be glutted and gorged, and not satisfied until it is. That jolly little camp fire is worth keeping an eye on, it is worth the trouble, even if we have to go half a mile to fetch it, to get a pail of water and ring the embers around with the wet so that the fire cannot spread. Never leave a camp fire burning; no registered guide would do such a thing, and no sportsman. It is only those who don't know or who are criminally careless who would. If the public will not take responsibility in this matter, the fire wardens are helpless. Some enemies these men must inevitably fight: the lightning which strikes a dead, punky stump in the midst of dry woods, which, smouldering a long while, finally bursts into flame; the spark from an engine; even spontaneous combustion due to imprisoned gases acted upon by sun-heat. But there is one enemy which the fire wardens should not need to meet, and that is man: the boy or girl camping, the man who drops a cigar stump or match carelessly onto dry leaves, the hunter who uses combustible wadding in his shotgun. Let us help the fire wardens, those men who live on lonely mountain summits or in the midst of the wilderness with eyes ever vigilant to detect the starting of a fire—let us help, I say, these fire wardens to get rid of one nuisance at least, and let us keep our great, cool, wonderful American forests as beautiful as they have ever been and should always be for those who are in a holiday humor.

CHAPTER VIII
OTHER SMOKE

There will not be much opportunity to dwell on all the wealth of information that comes to the real camper. The life of the woods is not only a lively one, but one teeming with intelligences and the kind of information which one can get no place else. My years of camping have stored my mind full of pictures and full of memories about which I could write indefinitely. In the practical activities of camp life we mustn't forget that the silent wonderful life of the wilderness is ours to study if we but bring keen eyes to it, quick hearing and receptive minds.

Let me tell you of one experience which I had some four years ago on the edge of a solitary little pond in the forest wilderness. Our way lay over a narrow trail, now through birches full of light, then through maples, past spruce and other trees, down, down, down toward the little pond which lay like a jewel at the bottom of a hollow. It was a favorite spot for beavers and we were going to watch them work. Their rising time is sundown, so we should be there before they were up. It was growing quieter and quieter in the ever-quiet woods, and when we hid ourselves behind some bushes near the edge of the pond on the opposite side from the beaver houses, there was scarcely a sound, and the drip of the water from a heron's wings as the bird mounted in flight, seemed astonishingly loud.

Soon the beavers, unaware of us, came out of their houses and began to work, steadily and silently. We knew them for what they were, builders of dams, of bridges, of houses, mighty in battle so that a single stroke from their broad flat tails kills a dog instantly, wood cutters, carriers of mud and stone— animals endowed with almost human intelligence and with an industry greater than human. And I never saw work done more quietly, efficiently and silently than I did that night by the edge of Beaver Pond.

As we sat there peering through the bushes I thought instinctively of the silent work which we do within ourselves or which is done for us. Deep down within us so much is going on of which "we," as we speak of the conscious outer self, are not aware. Take, for example, the frequent and common experience of forgetting a word or a name. Despite the greatest effort we cannot recall it, and finding ourselves helpless we dismiss the matter from our minds and go on to other things. Suddenly, without any seeming effort on our part the word has come to us. Now this reveals a great truth about a great silent power: *all we have to do is to set the right forces to work and frequently the work is done for us.* With this serviceable power within us, why not make use of it habitually? It renews itself constantly and waits for us to call upon it for protection, for comfort, for correction and strength. It insists only that we think as nearly rightly as we can. Beavers of silence are busy within us.

Much of the work of this silent power is done in our sleep-time. It is important, therefore, that our last thoughts at night and our first in the morning should be the best of which we are capable. Prayer is a profound acknowledgment of this power within us. We have all heard the expression, "the night brings counsel." And probably most of us have said, "Oh, well, we'll just sleep on that!" Why "sleep on it"? Because we have confidence in this silent power whose processes, whether we sleep or wake, are constantly at work within us, even as night and day, a natural power, directs the growth of tree and flower. Again we have counted upon the work of industrious beavers of silence—the silent workers within each one of us.

The woods are full of lessons never to be learned any place else. Insensibly are we, in this vast big intelligent life of the forest, led on to meditate about the things we see. I often wish not only that I could place myself at certain times in those solitary places by edge of pond, deep in forest, on the hillside, following the trail, but also that I might send a friend or two to the healing which can be found in the wilderness. For example, the girls who find nothing but troubles and vexations in life, who groan if the conversation languishes, are likely to have some of their troubles slip away from them and their talk become more cheerful. Who can be in the woods, who can live in the great out of doors and not feel optimistic, at least hopeful and interested? To every girl inclined to be moody, often to suffer from the conviction that living is difficult and perhaps not worth while, I commend camp life. Activity, distraction are its powerful and wholesome remedies for melancholy. In that life one is obliged to work mind and body much as the beavers work, one's attention is held to something every minute. The whole current of our thoughts has been changed and for the time being we are distracted from the old bruised ways of thinking. The very alteration that comes with wood life gives us a chance to think rightly. Who can be troubled or bored or bad tempered and follow the trail? Who can be indifferent and be conscious of the energy and intelligence of beaver and squirrel, of rabbit and bird, of deer and moose? Soon the whole misery-breeding brood of cares, of doubts, of perplexities that existed before we left our home drop away from us. We can use the influence of this vast sane life of the wilderness for ourselves and by its strength make good.

CHAPTER IX
FITTING UP THE CAMP FOR USE

Any girl who has crossed the ocean knows how impossible, the first time she entered her little white cabin, that bit of space looked as a place in which to sleep and to spend part of her time. There seemed to be no room in it for anything; it was difficult to turn around in, there were so few hooks on which to hang things, and the berth—dear me, that berth! So her thoughts ran. Yet gradually, as she learned the ropes, she was able to make it homelike. With experience she learned that the more bags she had in which to put things, the easier it was to keep this little stateroom in order. The next time she took with her every conceivable sort of bag for every conceivable sort of object. Also she had learned that the more she could do without unnecessary things in her cabin and steamer trunk, the more comfort was hers to enjoy. By the time she had crossed the ocean often, she had learned the art of having little but all that she needed with her—the art of making herself comfortable in a stateroom.

Even so is there an art in learning how to camp, a happy art of which there is always something left to learn. The oldest campers never get beyond the point where they can make a slight improvement in their kit or their methods. In the end you will work out your own salvation for the kind of camping you wish to do. It is my intention to point out to you only what might be called the ground plan of fitting up a camp for use. Those little individual adaptations which every one of us makes, increasing familiarity with camp life will help you to make for yourselves.

First, last, and always, when making out your camp lists, revise them carefully with the idea of cutting out everything unnecessary. All besides what you actually need will be clutter. The best way to do is to make out your lists, putting down everything that comes to you. Then go over them by yourselves and a second time with some one else. Your check lists for camp are important and should always be conscientiously made out, with nothing left to chance, nothing done hit or miss.

If you are to furnish a camp, remember that your packing boxes can do great work in helping to set you up in your new home. In rough camping such boxes do well for dressers, washstands and, with a little carpentry, also for clothes presses. A piece of enameled cloth on the top of the one to be used as a washstand, and a towel or white curtain strung on a string in front of it, behind which you can put dirty clothes, make a thoroughly satisfactory article of furniture. In camp there is no need to think about elegance. Fitness and usefulness are all the girl need ever consider. It is astonishing how much beauty your homely cabin and white tent will acquire—a beauty all their own.

For tent camping the usual camp cot bed is probably most satisfactory, for it is light and readily carried. If you are on the march and carrying at the most a tent fly for protection, you will, of course, sleep on bough beds or browse beds. Small, cut saplings, well trimmed, make good springs for beds. Any guide can help you to make the beds, and you would better be about it early, for it takes a good three-quarters of an hour to make a comfortable bough bed. Perhaps a few suggestions will not come amiss. You will, of course, have both good hunting knives, worn in a leather sheath on a leather belt, and belt-sheath hatchets. With the hatchet cut down a stout little balsam tree. From this break the tips from the big branches, having them about one foot in length. These foot-length stems make good bed springs and are the only bed springs you will have on a balsam couch unless you provide the spring yourself because of some green worm who is industriously measuring off the length of your nose, no doubt in amazement that there should be anything so extraordinarily long in the world. However, he is a harmless little chap, and the balsam tree having treated him very kindly, he will be greatly surprised at any other kind of entertainment which he may receive from you. Now, having got your "feathers," select a smooth piece of ground with a slight slope toward the foot. Press the stems of the feathers into the earth, laying them tier after tier as you have seen a roof shingled, until your bed is wide enough, long enough, and soft enough to give you a good and sweet-scented night of sleep upon it. Lay a fair-sized log along each side and across the foot. This balsam bough bed can be made up as often as you wish with fresh feathers. Place one blanket on top and it is ready for your use. If you have got pitch on your hands in doing this, rub them with a little butter or lard and it will come off.

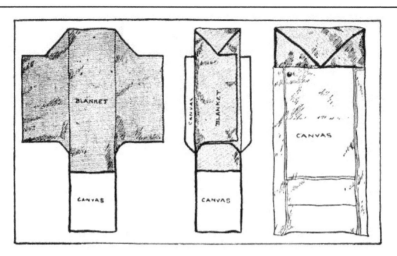

DR. CARRINGTON'S SLEEPING BAG.

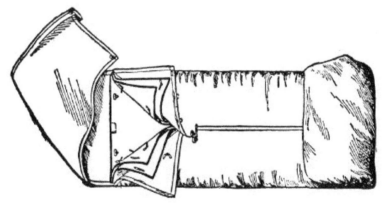

"KENWOOD" SLEEPING BAG.

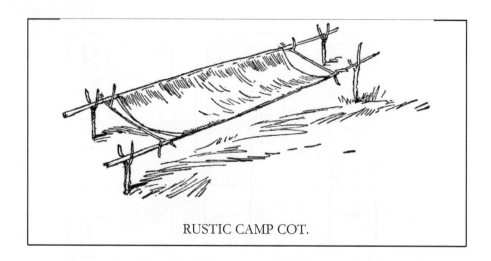

RUSTIC CAMP COT.

There is still an easier bed to make. A bag of stout bed ticking, filled with leaves and grass, forms an excellent mattress and has the virtue of being portable, for the bag can always be emptied, folded up, packed, and refilled at the next camp ground. A thin rubber blanket or poncho laid over this makes it an absolutely dry bed at all times. If you are to camp in a log cabin, probably the most comfortable bed for you to plan is a spring, bought at the nearest village, and nailed onto log posts a foot and a half high. With your ticking mattress filled with straw, your day lived in the great out of doors, no one will need to wish you pleasant slumber.

It is well to have a good supply of tarlatan on hand. This is finer than mosquito netting and therefore more impervious to stinging insects. If you camp in June, or the first week or so in July, you are likely in many parts of the country to find black flies, mosquitoes, and midges to battle against. There should be enough tarlatan to use over the camp bed and also enough to cover completely a hat with a brim and to fall down about the neck, where it can be tied under the collar. A more expensive head-net of black silk Brussels net can be made. This costs a good deal more, but the great advantage of it is, that the black does not alter the colors of the world out upon which one looks. Don't make any mistake about the importance of some kind of netting and fly dope, or "bug juice," as the antidotes for insect bites are sometimes called. There are various kinds of fly dope, any one of which is likely to prove useful. There is an excellent recipe for the making of your own fly dope in Breck's "Way of the Woods," which I give here.[6] A tiny vial of ammonia will also prove useful. One drop on a bite will often stop further poisoning from an insect sting. Inquiries should always be made beforehand whether one is likely to encounter black flies and midges. Those who have met them once are not likely to wish to have a second unprotected meeting. They are the pests of the woods and the wilderness.

[6] "Breck's Dope:

Pine tar	3 oz.
Olive oil	2 "
Oil pennyroyal	1 "
Citronella	1 "
Creosote	1 "
Camphor (pulverized)	1 "

Large tube carbolated vaseline.

Heat the tar and oil and add the other ingredients; simmer over slow fire until well mixed. The tar may be omitted if disliked."

I will give, just as they occur to me, a few other articles which will be useful in the camp life: a small cake of camphor to break over things in the knapsack and keep off crawlers; a small emergency box containing surgeon's plaster and the usual things; vaseline, witch hazel; jack knife; tool kit; a map of the region in which you are camping and a diary in which to take notes. To these might be added sewing articles, a sleeping bag if you care to use one, and a folding brown duck waterpail. The catalog from any sporting goods place will suggest a thousand other articles which you may care to have.

With a few planks to saw up into lengths, and a few white birch saplings, a most attractive camp dinner table can be made. Over this a piece of white oilcloth should be laid and kept clean by the use of a little sapolio. It is best not to buy an expensive stove for the cabin. A second-hand kitchen range, which can be purchased for a few dollars, will do quite well for the cooking cabin or shack, and an open Franklin stove for the living cabin. If one is going to camp in tents and wants a stove in one of them, it will be necessary to buy a regular tent stove. Anything else would not be safe.

As far as actual furniture is concerned, except for camp stools or benches and camp chairs, if you wish to be very elegant, the camp is now furnished. But there are still to be considered the necessary utensils for cooking and other purposes. I will enumerate them again just as they occur to me, and not necessarily in the order of their importance: kerosene oil can, molasses jug, pails, a tin baker, a teapot, tin and earthen dishes, tin and earthen cups, basins for washing, pans for baking and for milk, dishpans, dishmop, double boiler, broiler, knives, forks, teaspoons, tablespoons, mixing spoons, pepper box, salt shaker, nutmeg grater, flour sifter, can opener, frying pans—one with a long handle for use in cooking over open fires—butcher knife, bread knife, lantern, bucket, egg beater, potato masher, rolling pin, axe, hatchet, nails, hammer, toilet paper, woolen blankets, rubber blankets, crash for dish towels, yellow soap, some wire, twine, tacks, and a small fireless cooker if you know how to use one. A good fireless cooker can be built on the premises.

Possessed of these articles, any one who knows anything about the woods can be most comfortable. They can, of course, be added to indefinitely. One may make camp life as expensive and complicated as one pleases. But to do that seems a pity, for it is against the very good and spirit of the wilderness life. The wood life and all its new and invigorating experience should take us back to nature. It is for that we go into the wilderness and not to bring with us the luxuries of civilization. Part of the wholesomeness of camp life lies in learning to do without, in the fine simplicity which we are obliged to practice there. Common sense is the law of the wilderness life, and let us be sure that we follow that law.

CHAPTER X
THE POCKETBOOK

One of the objects of some girls on their camping expeditions is to keep the trip from becoming too expensive. The maximum of value must be got from the minimum of pence. And I think that is as it should be, for, with economy, the life is kept nearer a simple ideal, is made more active and more wholesome. All sorts and conditions of camping have been my lot, the five-dollar-a-day camping in a log cabin (?) equipped with running water and a porcelain tub, and the kind of camping one does under a fly with the rain and sunshine and wind driving in at their pleasure. Although I do not advise the latter as far as health results are concerned, given that the party is in fair condition they will be none the worse for the experiment.

Camping for a party of four or five should usually cost something between eight dollars and eighteen dollars apiece per week. This rate includes a guide and a good deal of service, a rowboat, a canoe, and no care about food. But the longer I camp the more I am of the opinion that the simpler and more independent the life is, the greater health and pleasure it will bring. It has been said about camping, "Much for little: much health, much good fellowship and good temper, much enjoyment of beauty—and all for little money and, rightly judged, for no trouble at all."

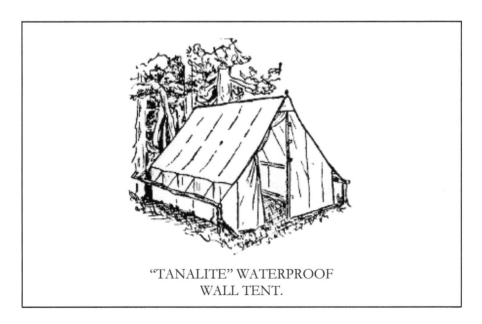

"TANALITE" WATERPROOF
WALL TENT.

TOILET TENT.

KHAKI STANDARD ARMY DUCK WALL TENT.

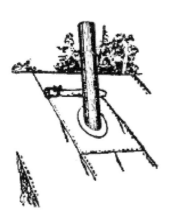

TENT STOVE-PIPE
HOLE.

FRAZER CANOE TENT.

WATERPROOF DINING FLYS FOR
WALL TENT.

The girl who is the right sort gets more fun out of camp life when she does at least part of the work herself. Let her economize and use her own ingenuity and do the work. Any group of three or four girls can provide all the necessary "grub" for themselves at $3 a week per capita. This sum does not include rental or purchase of tent. A good tent, 7 × 7, big enough for two at a pinch, can be bought complete (this does not include fly) for about $7. You can get tents second-hand often for a song, or as a loan, or you can rent your tent for 10 cents a day. Get at least a few numbers of one or several of the following sporting magazines: *Outing, Country Life in America, Forest and Stream, Field and Stream, Recreation, Rod and Gun in Canada*. Look in the advertisement pages of these magazines for the names of sporting goods houses and send for catalogs. Then choose your style of tent. The different kinds of tents are legion. The Kenyon Take-Down House, too, is a capital camp home. It is "skeet"-proof and fly-proof. Send to Michigan for a catalog, and then go like the classic turtle with your shell on your back. In groups of four or more, the $10 laid by for a vacation should bring two holiday weeks—possibly a day or so over; $15, three weeks and a bit over, and $20 a whole glorious month. Expensive camping may be the "style" in certain localities, but it is not necessarily the "fun."

For eight weeks this past summer my family of two members camped with two servants. In addition we had the occasional services of a man who did all the heavy work. There was not enough for the servants to do in the cottage and log cabin of our establishment. They were discontented, faultfinding, and wholly out of the spirit of camp life. All of the day that their tone of voice reached was helplessly ruined. The only way to keep the camp joy and pleasure was to keep out of their way. On our camp table we had silver, embroidered linen cloths, the same food, in almost the same variety, that we had it at home, and the same amount of service. All I can say is that it was a perfect nuisance—as perfectly planned and executed a nuisance as one could well conceive. Everywhere these servants looked they found things which did not suit them. What I think they wished was a modest twenty-thousand-dollar cottage in that great and wonderful wilderness.

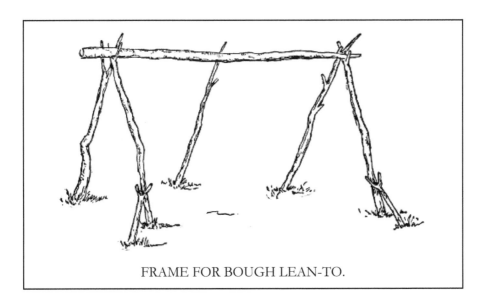

FRAME FOR BOUGH LEAN-TO.

BOUGH LEAN-TO.

In the autumn I camped alone for two weeks in a log cabin. I say alone. I was not alone, for I had three friends with me—a collie puppy, a blind fawn, and a year-old cat. They were the best of companions—for better I could not have asked. I never heard a word of faultfinding, and I was witness to a great deal of joy. It is a curious fact about camp life that if a girl has weak places in her character, if she is selfish or peevish or faultfinding or untidy, these weaknesses will all come out. But my four-footed friends were good nature itself, young, growing, happy, contented. And they had excellent appetites. I tell you this because I want you to see how much of an item their food was in the expenses I shall enumerate. This might be called a little intimate history of at least one camp pocketbook. The fawn had a quart of milk a day and much lettuce, together with the kind of food which deer live upon: leaves, grass, clover, ferns. I had to pay for her bedding of hay. The puppy and the cat shared another quart of milk between them. The cat hunted by night, but the puppy was fed entirely by hand on bread, milk, an occasional egg, cereals, and vegetables. My own fare consisted of all the bread and butter I wished, cocoa, condensed milk, bananas, apples, eggs, potatoes, beans, nuts, raisins, cauliflower, chocolate, and a few other articles. And there was, too, the denatured alcohol to be paid for—a heavy item, for I used only a chafing dish and a small spirit lamp. The milk was eight cents a quart on account of the carriage, the butter was thirty-eight cents a pound, the eggs twenty-five cents a dozen. Except for cutting up and splitting the wood for my open Franklin stove, the wood cost me nothing. But I paid a man a dollar for half a day's work. We weren't seven, but we were four in that camp community. How much do you think the food for all averaged per week in those two weeks? Three dollars a week, and we had all that we wanted and more, too.

When girls plan carefully and intelligently, when they exercise good sense in the cooking and care of food, there is no reason why, with a party of four or five girls, from three dollars to four dollars apiece per week should not cover all living, exclusive, of course, of the traveling expenses. And the camping can be done for less. I commend these expense items to all Vacation Bureaus and to Camp Fire Girls.

In the two weeks I camped alone I was very busy with my writing. To this I was obliged to give most of the daylight. Besides this, I had much business correspondence to attend to. It takes time to care properly for animals, and my pets had not only to be fed, but also to be brushed and generally cared for. I planned to spend some time every day with the blind fawn so that I might amuse her. I did all these things, took care of my little cabin, had time for a walk every afternoon, and went to bed when the birds did, to get up the next morning at five o'clock. Had I been able to give my thought entirely to the food question, I am certain that the expense of these items might have been made even less.

Some girls will think this is getting back to the simple life with a vengeance. So it was but I can assure you that those two weeks were most happy and profitable in every way—far better than the over-served, over-fed months which had preceded them. For any girl who needs to forget how superficial to the real needs of life the luxuries are; for any girl who is lazy in household ways; for any girl who needs character building; for any girl who is in need of deep breathing and the pines; for any girl who wants more active life than she gets in her own home; for any girl who is of an experimental or adventurous turn of mind; for any girl who needs to be drawn away from her books; for any girl who wants to form new friendships in a big, sane, and beautiful world where the greetings are all friendly; for any girl—for every girl—who wants much for little; the log cabin, the tent, the shack in the wilderness, by pond or lake, upon the hillsides or in the valleys, will prove a "joy forever."

CHAPTER XI
THE CAMP DOG

When I began to go into the wilderness to camp, I was much more credulous than I am now. Everywhere I went in the woods I saw an implement which looked like a cross between a pickaxe with a long handle and the largest pair of tweezers ever seen. This was always lying up against something as if just ready for use, much as one sees an axe resting against a cabin wall or on a chopping block. I couldn't make out what this could be used for. Finally, curiosity getting the better of me and no opportunity for seeing it used offering itself, I asked.

"Oh, that," answered the guide with a twinkle in his eye, "that is the camp dog."

"How nice!" I thought. "Why is it called camp dog?"

"Well, you see it does most of the work for us and being so faithful and handy we've just got naturally into the way of calling it a camp dog."

I was still more impressed when he gave me then and there several illustrations of its usefulness. But the end of the tale of the camp dog is not yet,—in fact it was a very long tale for me, the end of which you shall have in good season.

Generally speaking it may be said that it is the guide and not this implement which is the camp dog. It is he who is faithful, always handy, always willing. And it is he who is more imposed upon than any other member of the camp community. The guide is a responsible person,—*the* responsible person. He is usually registered and his pay is always good. He needs every dollar he gets and every bit of authority, too, for he works hard and often for groups of people who are thorough in only one respect and that is in their irresponsibility. The guide has to be sure that fires are kindled in the right places and that they are really out when they should be; he must keep his party from foolhardy acts of any kind; he must be sure that they have a good time and certain that they are not overtaxed; if it comes off cold or is cold, he must keep them warm; he must see, despite every vicissitude, that they are enjoying themselves; he must do the cooking—and he must be a good cook,—boil the coffee, wash the dishes, pitch and strike the tents; he must pilot the members of the party to the best places for fishing, often bait their hooks or teach them how to bait, dig their worms; and give their first lessons in casting a fly; must instruct them in all necessary wood craft and keep them from shooting wildly; he must see that the game laws of the state are observed, also the fire laws; if anything should happen to a member of his party, he will, in all likelihood, be held responsible for it; and finally, always and all the time, no matter how he himself feels, he must be agreeable, obliging, useful.

Now if the man who has all these burdens to bear is not a camp dog, I should like to know what he is? To those of us who have been into the woods year after year, it is a sort of boundless irritation to see some members of the camping party sitting about idle while the guide does the work. Part of the value of camp life is its activity, its activities. Another part of its good is the skill which comes from learning to be useful in the woods. The life out-of-doors should be a constant training in manual work,—call it wood work if you wish. I am reminded of a story told in "Vanity Fair" about a lazy, indifferent student who was in the class of a famous physicist. The freshman sprawled in the rear seat and was sleeping or was about to go to sleep.

"Mr. Fraser," said the physicist sharply, "you may recite."

Fraser opened his eyes but he did not change his somnolent pose.

"Mr. Fraser, what is work?"

"Everything is work."

"What, everything is work?"

"Yes, sir."

"Then I take it you would like the class to believe that this desk is work?"

"Yes, sir," wearily, "wood work."

From the moment that school of the woods is entered every girl has her wood work cut out for her, if she is taking camping in the right spirit. It is all team play in the wilderness, or if it is not, it is a rather poor game. Helpfulness is one of the first rules and every camper should be willing to help the guide. Usually the guides are a fine set of self respecting, dignified, resourceful men. And I think it might be said with considerable truthfulness that when they are not what they ought to be, it is nine times out of ten due to the undesirable influence of the parties they have worked for. Your guide is your equal in most respects and your superior in others. He should be met on a footing of equality. I use this word advisedly and I do *not* mean familiarity. Well-bred girls do not meet anyone, whether in the wilderness or in civilization, on this footing immediately. The party should be willing and glad to help the guide in every possible way. That does not signify doing his work for him but it does indicate helping him.

A routine of some sort should be adopted and is one of the best ways to assist him. One girl should be on duty at one time and another at another and all in regular rotation. No camp life can go on successfully without some law and order of this sort. For it is just as necessary for the smooth running of household wheels in the log cabin as it is in the city home. Whoever occupies the guide's position, that is the one who is chiefly responsible for everything, should be ably helped by the whole party but not by the whole party at the same time. Evolve a system for the particular conditions of the camp life in which you find yourself and stick to it. Let one girl or one set of girls help one day and another the next. Let the girl be detailed to do one kind of work one day and another another. This system, with proper rotation, means that nobody gets tired of her work. A girl cannot be too self-reliant if she is ever to be wise in the way of the woods. There is no need for discouragement if everything is not learned at once, for camping is like skating and is an art to be learned only through many tumbles and mistakes. Be prepared to take it and yourself lightly—in short, to laugh readily over the mistakes made in the art of living in the woods.

Now we have come to the very tip of the tail of the camp dog. You will be interested to know how an old timer was obliged to laugh at herself. I am ashamed to tell you how recently this occurred. I was in the northernmost wilderness of the state of Maine, and near a big lumber camp, when I saw a "camp dog" lying on the ground, its long axe handle shining from use, its pickaxe blade a bright steel color, and the tooth at the back looking as if it had been often used. I was delighted.

"Oh," I said to my guide, "look at that camp dog lying there!"

He was particularly attentive to my pronunciation, for he said I pronounced some words, such as "girl," as he had never heard them pronounced before. I saw a curious expression pass across his face.

"What did you say that was?" he asked.

"Why, that camp dog lying there."

"Camp dog!"

Then he began to laugh and he kept right on until the woods echoed with his roars.

"Well," he said finally, wiping away the tears, "if that doesn't beat everything! That isn't a camp dog, that's a cant dog,—you know what you cant logs and heavy things over with, roll 'em over and pry 'em up with when you couldn't do it any other way. My grief, to think of your calling that a camp dog all these years!"

And he went off into another guffaw.

CHAPTER XII
THE OUTDOOR TRAINING SCHOOL

Many girls think of outdoor life as of something to be enjoyed if they have plenty of time. As a matter of course they take their daily bath. But the outdoor exercise comes as an accessory. It is still unfortunately true that boys more than girls take camp life for granted. Yet girls, and students particularly, should realize that it is economy of time to be out of doors. This they need both for their work and for their health. Outdoor exercise, with its bath of fresh air and the natural bath of freshly circulating blood it brings with it, its training school for the whole girl, is as essential as the tub or sponge bath. But how many of us think of it in that way?

To be outdoors is to have the nerves keyed to the proper pitch. If fresh air is not a tonic to the nerves, then why is it that moodiness and depression fall away as we walk or row or lie under the trees, and we become saner and more serene? When one is depressed the best thing to do is to go out of doors. Altogether aside from any formal wisdom of book or student or teacher, there is wisdom with nature. *If the head is tired, go out of doors! If the body is fagged, go out of doors! If the heart is troubled, go out of doors!* The life out there, as no life indoors can, will make for health, for charity, for bigness. Petty things fall away, and with nature equanimity and poise are found again. It isn't necessary to bother someone about woes real or imaginary. All that is necessary is to get out among the trees and flowers, the sky and clouds, the joyous birds and little creatures of field and wood, and hear what they have to say. There will be no complaining among them, even about very real difficulties.

A great deal is heard concerning hygiene in these days, the study of it, the practice of it. The biggest university of hygiene in the world is not within houses but outside, up that hillside where the trees are blowing, in the doorway of our tent, on the lawn in front of the house, out on the lake, even on a city house-top, and, last resort if necessary, by an open window. One reason why many people are concerned about this question of hygiene is because they know that not only are human beings happier when they are well and strong, but also because a healthy person is, nine times out of ten, more moral than one who is sick or sickly. Ill health means offense of some kind, often one's own, against the laws of nature or society. We have, too, to pay for one another's faults. But life lived on sound physical principles, with plenty of sunshine, cold water, exercise, wind, rain, simple food and sensible clothing, is not likely to be sickly, useless or burdensome.

BITTERN

LOON

PARTRIDGE

RED-BREASTED MERGANSER

WOODCOCK

MALLARD

The body is not a mechanism to be disregarded, but an exquisitely made machine to be exquisitely cared for. Nobody would trust an engineer to run an engine he knows nothing about. Yet most of us are running our engines without any knowledge of the machinery. Why should we excuse ourselves for lack of knowledge and care when, for the same reasons a chauffeur, for example, would be immediately dismissed? How many of us know that the nerves are more or less dependent upon the muscles for their tone? How many of us realize how important it is to keep in perfect muscular condition? We sit hour after hour in our chairs, all our muscles relaxed, bending over books, and begrudge one hour—it ought to be three or four!—out of doors. The person who can run furthest and swiftest is the one with the strongest heart. The person who can work longest and to the greatest advantage is the one who has kept his bodily health.... *It may be laid down as an absolute rule that any individual can do more and better work when he is well than when he is not in good physical condition.* Ceaseless activity is the law of nature and the body that is resolutely active does not grow old as rapidly as the one that is physically indolent.

Much out-of-door life, much camping, keep one young in heart, too. It isn't possible to grow old or sophisticated among such a wealth of joyous, wholesome friendships as may be found in nature, where no unclean word is ever heard and where no unfriendliness, no false pride, no jealousy can exist. A great English poet, William Wordsworth, has told us more of the shaping power of nature, its quickening spirit, its power of restoration, than any other poet. It would be well for every girl to take that wonderful poem "Tintern Abbey" out of doors and read it there. Wordsworth, still a very young man when he wrote it, tells how he loved the Welsh landscape and the tranquil restoration it had brought him

"'mid the din
Of towns and cities."

A higher gift he acknowledges, too, when through the harmony and joy of nature he had been led to see deeply "into the life of things."

There is something the matter with a girl who hasn't an appetite, as sharp as hunger, to escape from her books and camp out of doors. If outdoor life cannot engross her wholly at times, banishing all thoughts of work, then she should make an effort to forget books and everything connected with them for a while. A young girl ought to be skillful in all sorts of outdoor accomplishments, rowing, swimming, riding and driving if possible, canoeing, skating, sailing a boat, fishing, hunting, mountain climbing.

Fortunately there is more of the play-spirit connected with outdoor life than there used to be. Both school and college have fostered this wholesome attitude. If a girl doesn't like active sports she should cultivate a love for them. You can always trust a person who is accomplished in physical ways, for anyone who has led an intelligent out-of-door life is more self-reliant. Her faculty for doing things, her inventiveness, her poise, her "nerve" are all strengthened. I recall an instance of this "faculty" and inventiveness. We were on a wild Maine lake when an accident happened to the canoe, a necessity to our return, for we were far away from all sources of help. Apparently there was nothing with which to mend it. But our Indian guide found there everything he needed ready for his use. He scraped gum off a tree, he cut a piece of bark, and then he rummaged about until he discovered an old wire. With these things he securely mended a big hole. Oftentimes it seems as if the very appliances with which city children are provided tend to make them incapable.

YELLOWBIRD

FIELD SPARROW

GOLDEN-CROWNED THRUSH

SONG SPARROW

CHIPPING SPARROW

WOOD THRUSH

HERMIT THRUSH

SWAINSON'S THRUSH

WILSON'S THRUSH

PHŒBE BIRD

SCARLET TANAGER

MARYLAND YELLOWTHROAT

BLUEBIRD

WREN

BLUE JAY

CHICKADEE

RUBYTHROAT

WHIP-POOR-WILL

NIGHT HAWK

SCREECH OWL

The girl who lives out of doors acquires unlimited resourcefulness. Outdoor life quickens and sharpens the perception. And for the girl to have her power of observation sharpened is worth a great deal. The capacity for accurate and quick observation education from books does not always develop. One must go back to nature for that, one must live out in the woods and fields all one can, one must be able to tell the scent of honeysuckle from the scent of the rose, and know the fragrance of milkweed even before that homely weed is seen, and know spruce, balsam and white pine even as one knows a friend. Eyes must be able to detect the differences not only in colors and shapes of birds, but in their flight, and ears know every song of wood and field. Then the services of beauty, its music, its color, its form, will be always about us and nature's health and strength and beauty become our own, not only her gaiety and "vital feelings of delight," but also her restraint upon weakness, and her kindling to the highest life—the life that is spiritual.

BLACK SPRUCE

BLACK OAK

BIRCHES

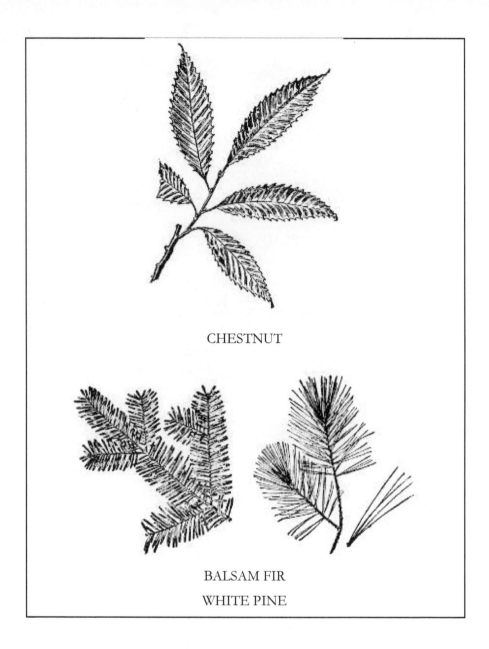

CHESTNUT

BALSAM FIR

WHITE PINE

BEECH

LARCH

HORSE CHESTNUT

MOUNTAIN MAPLE

CHAPTER XIII
THE CAMP HABIT

If there were no such thing as habit, life would be nothing but a perpetual beginning and recommencing over and over again. All that we do or think marks us with its imprint, leaving behind it a tendency—a tendency towards repetition is the beginning of habit, and because of it we can get the camp habit just as we can get any other habit. The instinct to repeat our camping out of doors gradually grows stronger. At last, scarcely conscious of the existence of the demand, we have come to feel that we cannot pass our holiday in any other way. The first camping experience stands out in bold relief because it is new. As we live into it, its first impressions are lost. And it is at this moment, if we are made of the right stuff and have in us the right longings and needs, that we begin to have the camp habit.

Just as with people, maybe we scarcely realize how much it means to us. But let us stop to think about it, let us give this good camp habit a full opportunity if we can in our lives. Already the camp habit has become a need, almost an imperious demand. We feel that once in so often it must be satisfied and in the splendid grip of this good habit we make way for it. Never let us become dull to any of its values. Never let us forget, however shot with black and white it may be, even gray at times, the difficulties of camping may make life seem—never let us forget the treasures that it pours in upon us and the ways in which the camp habit serves us.

It is a sad and a great truth which perhaps women and girls have not yet fully realized, that the whole manner of our body, of our souls is controlled by the goodness, or the badness of our habits, our moral character, our physical temperament. There is a sort of natural medicine, raising what is not good inevitably up to what is better. That is what the camp habit does for us, raising what is not healthy, not strong, not sane, not joyous, not self-reliant up to what is strong, healthy, joyous and full of self-control. Is not this alone sufficient reason for giving the camp habit once in so often full sway in our lives? What better could we do than, in order to re-establish ourselves, to claim again the wise big relationships of out-of-doors and a thousand and one little and big friends whom we can find there?

Bad habits are thieves, for they take away our energies, our abilities, our joys. And the indoor habit is a thief. It shortens life, it takes away from health, it saps energies, it dilutes joys, it makes foggy heads and punky morals. The sane girl will get out of doors every opportunity instead of spending her time in a hot room, playing cards, or eating stuff that is not fit to put into the human stomach or flirting with boys, who if they are the right sort of boys, would much prefer, too, to be out of doors. Good habits, like this camp habit are benefactors, great philanthropists; they strengthen us and they give us more energy. They increase our ability, they multiply our joys compound interest-wise. Good habits are careful accountants and every day or every year as it may be, they put the interest of strength, of intelligence, of joy, in our hands to be used as we think best. The camp habit wisely used, obliges us to open our eyes and see life more truly. It obliges us to lift our own weight, take our part in things, that part may be washing dishes or it may be turning griddle cakes,—it forces us to know ourselves better and it gives us more power to control ourselves. The camp habit—get it quickly if you haven't it already—assures us of good health and success where, for example, the indoor habit has brought us nothing but ill health and failure. It is a habit worth while getting, isn't it?

A good many of us know ourselves, such as we are, pretty well and we feel that we do not want to know ourselves any better. Things are bad enough as they are. Yet if we can't have a more intimate knowledge of ourselves, if we don't arrange our lives better, if we don't plan for the future more carefully, what are our lives likely to be like when the curtain goes down? How are we ever going to take the proverbial ounce of prevention if we are not certain to a fraction what it is we must prevent? Camp is a splendid opportunity to think a little about those things of which we have been afraid to think. It is a good opportunity to meditate, a friendly world to which to go to know ourselves better. It is an old saying that the first step towards the recovery of health is to know yourself ill. In that great out-of-door world which our American camp life represents it is easier to find ourselves morally than it is indoors, we get more help for one thing. It is almost an instinct in great trouble or bewilderment or difficulty to escape into the out-of-door world, to get back to earth and to ask from the great mother those counsels we hear dimly or indifferently indoors.

Wisdom will not be found in one camp holiday or in fifty or in a lifetime even. But it is rather strange, isn't it, that the person whom we know least is so frequently ourselves? We know very well that the most learned man or woman is not the one whose head is stuffed with information, is not necessarily the conspicuous or famous man or woman, but is, rather, the human being who knows himself. And this human being may be not our teacher, but our janitor or a nurse who takes care of the baby or that fellow who seems so simple, the guide who has our camping trip in charge. Indeed, there is scarcely a class of men who seem in better control of themselves and who have a better working knowledge of themselves and others than the highest type of guide. All the associations of that great out-of-door life, its demands, its privations, its sudden needs, its great silence, its dumb creatures, its wonderful beauty, have taught the man of the woods a wisdom no school, no university, can offer merely through its curriculum. We can't realize too early how well worth while that wisdom is for every girl to have. Not a thing of book learning, but a power that makes one truthful with oneself, eager to acknowledge what is bad and to change it. Frank, courageous, tried in commonplace wisdom, and with a knowledge of other human beings.

There is one kind of idea—and it is worth while meditating in the woods on the leverage power of even one very little idea—that can always be found out of doors. I mean a healthful idea, the kind of thought that makes us stand straighter, that strengthens the muscles of our backbone, that makes us act as if we were what we wish to be. There is no other force in the world that can so readily straighten out a crooked boy or a crooked girl as this same Dr. Dame Nature.

———

CHAPTER XIV
OTHER CLEANLINESS

Clean? Of course, we all know what cleanliness means. It is not possible to drive, to ride in a trolley, to go on a train without being impressed with at least the advertising energy that is put into trying to get or keep the world clean. Dear me, there are the ever-present, cheerful Gold Dust Twins, well up with the times, you may believe, and nowadays taking to aviation. Their aeroplanes may not be very large, but they are clean as gold dust can make them, and the twins, without any of the friction that comes from dirt, are flying at last. What's more, intrepid as some old Forty-Niner, they are penetrating the camper's wilderness. Most of us do not want to be twins, and we certainly do not want to be gold dusters or any other kind of dusters, yet we should miss these jolly little youngsters. And there are Sapolio and Sunny Monday advertisements and Pears' soap—have you used it?—and a dozen other kinds and goodness knows what not besides.

Yes, we Americans, and especially American women in the household, know what it is to make an effort in the midst of heated, dusty or uncared for streets to keep our houses and everything in them clean. In Pennsylvania you see the people scrubbing off white marble steps. In New England they turn the hose on the outside of their white farm houses. In the West they flood the side-walks to keep the dust and heat down. And our houses? Well, all houses are being built with bath tubs nowadays, even our camps, which is more than can be said for very good houses indeed in other countries than America. Some people think that camping is an excuse to be dirty. Often they are very nice people, too, but they keep a dirty camp. They don't keep even themselves clean.

But there is another kind of cleanliness, not superficial, not that of the skin, or of the clothes or of the cabin, about which we are coming to think more and more deeply. It is what might be called vital cleanliness, the cleanness of stomachs, of the intestines, of all the vital organs. We begin to realize the truth of what those most helpful of missionaries, the health culturists, are saying: One may be clean superficially, that is one may scrub enough and yet vitally be very far from clean. We know, although it is of the greatest assistance to keep the skin free and vigorous so that it is able to do its part of the house-cleaning work for our systems, that vital cleanliness, clean, strong, internal organs performing their work with the vigor of well-constructed engines, uninjured by foolish clothing, unharmed by impure food, keen for opportunity to grow and be vigorous—we know, I say that that cleanliness is more important than skin cleanliness. Indeed, without such deep-seated cleanliness it is impossible for the skin to be really clean.

But clean how? I wonder whether we are clean in the way I mean. Yes, we are clean in our houses, perhaps in our camps, clean on the outsides of our bodies, clean probably, on the inside. Yet no one of these kinds of cleanliness is what I have in mind. Can any girl by the camp fire guess what it is? I will not say it is more important than household cleanliness, although it is so,— vastly more so. I will not say that it is more important than bodily cleanliness, external and internal, yet it is so,—vastly more so. I could almost say that it is more important than anything else in the world of human experience. Do you know what it is now? *It is cleanness of the mind, cleanness of the soul,* and of that kind of purity the great outdoor world is one indivisible whole.

On this cleanliness of mind and soul all the vital activities of the day depend, all the growth, the gain, the development. It might be well said that the way we take up the sun into our bodies—and we could not live any length of time without some sun—depends upon the cleanness or uncleanness of this mind and soul of ours. What we shall eat, what we shall hear, what we shall see, what we shall look forward to, what we shall care for—all these things will be according to laws as inevitable as those governing the sun and moon and stars, valuable or worthless, vicious or sacred, as we feel them and we make them. We dip our fingers in pitch and pick up a book. What is the result? Any child could tell us that we ruin the book with our pitch-covered fingers. We dip our minds into filth, a nasty story, a perverted way of looking at things which in themselves are good and of God's plan, or we actually commit some ugly act ourselves and then we go out into the presence of those things which are clean, the sunshine, the hills, the lakes, the woods, the white lives of others, the ideals which, it may be, have been ours. Do you suppose we feel or see that sunshine, or that we are aware any longer of the white lives of others, that our past ideals are evident to us when our hearts and minds are no longer clean? Do you suppose that there is anything in nature which comes home to us in quite the beautiful way it once did, the flowers, the birds, the song of the wind, the little creatures of the wood? Can they ever be entirely the same? No, by an inevitable law of compensations some of the fullness of our joy in these things is gone. If we want to be really happy it does not pay to think evil, to touch evil or to commit it.

When our hands are dirty we know it, and if we have been careless about them we are ashamed. If people's bodies or camps are not clean it is painfully easy to know that, too. But a dirty mind, who could ever tell anyway that we had one? Who could ever tell? I will tell you: *Every one knows it,* or perhaps, better, every one feels it. If we are not good, if our minds are not clean, our presence in some mysterious way proclaims that fact. If we have injured some one, if we have been foul-tongued, others will know it with no need for any one to tell them. Even the little rabbit we meet in the woods will not greet us in so friendly a way. *We need not think that because we are concealing a bad thought that it is therefore hidden.* No, indeed, it is screaming away like some ugly black crow on a spruce tip, and there is no one within hearing distance who, whether he wishes to or not, does not hear what it says.

The mind has its plague spots even as the body, and one has to work—because of one's environment or some inheritance which has made us not quite wholesome by nature, or because of friends whose feelings one would not injure, and yet who are not what they ought to be,—one has often to work to keep the mind clean. But as you would flee from the plague, run from a dirty story. Don't let the camp life be spoiled by anything to be regretted! Do not let any one touch you with it, even with a word of it. Keep a thousand miles away if you can from folk who have an impure way of looking at life, and camp is a good place to get away from such people. Shut your minds against them. One is never called upon on the score of duty to have an unclean mind because others have it. And if through some misfortune, something that is unlovely, unclean, has been impressed upon you, fight valiantly not to think of it, to put it away from you. And never forget that to rule our spirits, to be in command of our minds, to have them wholesome and sweet and clean as a freshly swept log cabin, is greater than to win such victories as have come down in the records of history.

I remember that when I was a child, I thought my heart was white and that every time I said or thought anything naughty, I got a black spot on its surface. I dare say that in the first place some dear old negro woman put this fable into my mind. And, dear me, some days it seemed to me that heart of mine was more spotted than any tiger lily that ever grew in any neglected garden. Perhaps it was foolish to think such a thing. I do not know, I only know that there were times when I was mighty careful of that white heart of mine,—wrapping it up in a pocket handkerchief would not have satisfied my eagerness to keep it clean. And what better could one wish than to go on one's holiday, and on forever, with the white shining heart of a child?

———————

CHAPTER XV
WOOD CULTURE AND CAMP HEALTH

It is far better for the girl to be out in a wilderness world which demands all the attention of both heart and mind, than to be leading an idle or sedentary life at home. If there is one word which above all others expresses the life of the woods, it is the word WHOLESOME. It is a normal, active, "hard-pan" life which takes the softness not only out of the muscles, but also out of the thoughts and the feelings. It tightens up the tendons of our bodies and the even more wonderful tendons of the mind.

Often, to paraphrase Guts Muths, a girl is weak because it does not occur to her that she can be strong. She fails to lay the foundations of health and strength which should be laid; she fails to make the most of the energy that she has; she fails to think of the future and how important in every way it is that she should be robust and full of an abounding vitality. It is a matter of the greatest importance to the world spiritually, morally, physically, that its girls should be strong. To be out of doors insures abundant well-being as nothing else can. The wilderness instinct, the instinct for camping and all its out-of-door life and sports, is the healthiest, sanest, and most compound-interest-paying investment a girl can make.

But by an intelligent approach to this life, more can be put into it and therefore more can be taken out, than by some blindfolded dive into its mysteries. To know how to do a thing worth doing and to do it well, is both wise and economical. Some of the physical aspects of our life will give all the more value because of the payment of an added attention. A few simple rules for the physical side of camp life will do quite as much for the body as an orderly routine can do for the camp housekeeping.

Simply because you are in camp, never do anything by eating or drinking or over-strain or folly of any sort, that is against the law of health. To break the laws of health is as much a sin in camp as out of it.

Eat an abundance of simple, wholesome foods, using as much cereals, fruits, and vegetables as you can get. Don't neglect the care of your teeth merely because you are in camp.

Do not drink tea or coffee. Stimulants are unnatural and unwholesome; no girl and no woman should ever touch them. If you have begun to drink tea and coffee, camp is the place to give them up once and for all time. The sooner the better.

If you can get a cool bath in stream or pond and a rub down with a rough towel, so much the better. Exercise both before and after the bath, and be sure, by rub down and exercise, to get into a good glow. The rub down is of especial importance, for it stimulates all the tiny surface veins, is gymnastics to the skin, and frees the pores of any poisonous accumulations which they may be holding. Drink a glass or two of pure water when you get up and the same between meals.

Never wear anything tight in camp or elsewhere. Within the circle of the waist line are vital organs which need every deep breath you can take, every ounce of freely flowing blood you can bring to them, every particle of room to grow you can give them. The Chinese woman who cramps her feet sins less than we who cramp our waists.

Sleep ten or eleven hours every night.

Study to make your body well, strong, and useful.

If you do all these things, you need not worry about beauty; you will possess what is of infinitely more value than a pretty face and abundant hair, in having a sound, wholesome body, self-controlled, instinct with joy, with clean, glowing skin, a pleasure to yourself and to everybody else. Clear vital thoughts and a keener spiritual life will both be yours. Because of the days in the woods it will be easier to be good, easier to be happy, easier to do the brain work of school and college.

Part of the title of this chapter is Wood Culture. I have something in mind that is more than physical culture: The wilderness cure, the lesson of the woods, a high spiritual as well as physical truth. For the girl who keeps her eyes open, here are forces at work, mysterious, inspiring, wonderful, that awake in her all the dormant worship and vision of her nature. Yet of physical culture in these weeks and days in the woods too much cannot be said, for, as the world is beginning to realize, on one's physical health, cleanness, sanity, rests much of that close-builded wonderful palace of mind and soul. Every squad of girl campers should have its physical culture drill, its definite exercises, taken at a definite time, for ten or fifteen minutes. Ten or fifteen minutes are probably all that are necessary when practically the remainder of the day is spent in camp sports, canoeing, fishing, climbing, hunting and so on. The object of these physical exercises should be all-around development; the drill should be sharp and light with especial attention paid to breathing and to the standing position. A steady unflagging effort should be made to correct round shoulders, flat chests, drooping necks, and bad positions generally. Many and varied are the exercises taught in school and college,— exercises to which all girls have access. I make no apologies for suggesting a few of the simplest by means of which any squad of girl campers can make a beginning in physical culture.

(1) From attention (hands on hips), place the palms of the hands flat on the ground, keeping knees straight. Then bring arms up above head. Do this eight times.

(2) With hands on the hips and the hips as a socket, rotate the whole trunk first five times in one direction, then five times in the opposite, being sure that the head follows the line of the rotating trunk. The difficulty of this exercise can be increased by placing hands clasped behind the head, and then later over the head. But the exercise should be undertaken first with the hands on the hips.

(3) In between each exercise take deep breathing for a few seconds, rising on the toes as you inhale and lowering as you exhale.

(4) Stand with the feet apart and arms horizontal. Without bending the knee place the right fist on the ground next to the instep of your left foot. Then raise the body and reverse, placing the left fist on the ground next to the right instep.

(5) After this some free exercises with the arms, taken with the head well up, chest out, and shoulders back, make a good, sharp light finale.

These exercises repeated several times make an excellent beginning for any day, either in or out of camp. You may unfortunately be going through a state of mind, when clean skin, good lungs and digestion, seem to you negligible factors in life. How tragically important these factors are, be sure you do not realize *too* late, when both body and soul, health and morals, have been undermined.

Most girls need to look upon camp life as an incomparably rich opportunity to gain in an all-round physical development. The life itself, aside from its possible physical culture exercises and its sports of rowing, paddling, swimming, climbing and walking, is the big architect of a splendid substructure for health. By taking thought, refusing to eat greasy, unwholesome food, getting plenty of sleep, avoiding over-strain, taking corrective exercises, cool baths and rub downs, there is no better health builder than the wilderness life. A wise Danish man said that "He who does not take care of his body, neglects it, and thereby sins against nature; she knows no forgiveness of sin, but revenges herself with mathematical certainty." In the woods nature keeps reminding you of this fact, and you are never allowed to forget it for any length of time.

It is only sensible to care for one's health. It is not necessarily old maidish or silly to take precautions that the camp health should be at its zenith all the time. No one would think of criticising a man for being particularly careful of his horses under new conditions. This is precisely what we should be for ourselves. Your thorough-paced sportsman is always regardful of his physical condition. I have spoken about the drinking of pure water, the care of food, the folly of taking great risks, and of other details. There are more factors, as well, which will be at work in obtaining and maintaining good health conditions.

The right sort of underclothing—and women seldom wear suitable underwear—should be worn. It should be high necked, with shoulder caps and knee caps, and should be of linen mesh. Every girl who is in fit condition should see that each day has a brief period at least of hard, warm, strenuous work in it. A sweat once a day, with a proper rub down afterwards, is one of the best health makers on record. In "By the sweat of thy brow shalt thou labor" was enunciated one of the greatest of natural laws. If it were possible for each one of us to sweat once a day, we should scarcely ever know what sickness is. But our over-refined civilization makes even the use of the word an offence to certain middle class people who care more for the so-called propriety (they are the folk who say "soiled" handkerchief instead of dirty, and "stomach" when they mean belly, and yet are ready to use such a detestably vulgar word, straight out of the mouths of the lowest classes of immigrants, as "spiel") of what is said than for its truth and strength. Lay it down, then, that one of the first of the camp health rules is a sweating every day. Third among the camp rules is to keep the bowels open. Do you know what one of Abraham Lincoln's mottoes for life was? "Fear God and keep your bowels open," and in this saying there is no irreverence whatsoever, nor any sacrilege, but only a profound common sense that is a credit both to the Maker and the great man who spoke the words. Cascara is the best and safest laxative for a girl to use in camp. It should be bought in the purest tablets or liquid form on the market, and all patent cascara nostrums should be avoided.[7]

[7] If there is a privy in the camp great care should be taken that, for every reason, it is placed at a sufficient distance from cabins and tents. It should *not* be placed on a slope that could possibly drain off into any water supply. An abundance of ashes should always be kept within the privy and no water of any kind be poured into the box. A few cans of chloride of lime should, if possible, be kept on hand; and one can opened and in use in the closet. Chambers and slop pails should not be emptied in the immediate vicinity of the cabins but at some distance and in different localities. There is no greater abomination on the face of the earth than a dirty camp, and no place which so thoroughly tests one's love of order, decency and cleanliness. If you are following the trail and go into "stocked" camps for the night, shake and air the blankets thoroughly, and, out of courtesy to those who will follow you in their use, shake and air the blankets when you get out of them in the morning.

If a girl is delicate or under the weather in any way, she must take more than the ordinary care of herself or she may have a head-on collision with out-and-out illness. The new mode of living, the various kinds of exposure—especially to wet weather—, the larger quantities of food eaten because of an appetite stimulated by the vigorous outdoor life, the temptation to overdoing—all these possibilities should be kept in mind and avoided as dangers. Don't be silly about overdoing. Harden yourself slowly for the life; avoid competition. It is far better to have lived your camp life successfully and to have come out of it fresh and vigorous, than it is to have done a few "stunts" and have come out of it fagged, overstrained and ill. It is well the first days of camp life to try to eat less than you want; by this act of self-control you will avoid the plague of constipation which follows so many campers. Moderate eating will mean more sleep, too. Abundant water drinking and a few grains of cascara should be able to remedy all the ills to which camp flesh is heir.

As a girl takes thought about this care and culture of the body, making herself clean within and without, higher lessons and perfections, both of the mind and of the soul will come to her as inevitably as the earth answers to the touch of rain and sun. Do you want to be happy? Very well then, learn in the woods to be well, consider the laws of health, and remember first, last, and always that good health, not money or position or fame or any shallow beauty of feature, is the greatest and soundest security for happiness.

———

CHAPTER XVI
WILDERNESS SILENCE

Most friendships among girls, and older people, too, suggest that if there is one thing which is hated, it is silence. If silence does happen to get in among us in camp, how uneasy we are! After an awkward pause we all begin to talk at once,—any, every topic will serve to break the hush which has fallen upon us. And if we don't succeed in getting rid of this silence—something apparently to be regarded as unfriendly and ominous—we make excuse to do something and do it.

But of silence Maurice Maeterlinck, the great Belgian author of "The Bluebird" and of many other plays, too, says that we talk only in the hours in which we do not live or do not wish to know our friends or feel ourselves at a great distance from reality. But where do we live more truly than in our camp life? Then he goes on to say what I think is equally true: That we are very jealous of silence, for even the most imprudent among us will not be silent with the first comer, some instinct telling us that it is dangerous to be silent with one whom we do not wish to know or for whom we do not care or do not trust.

Let us admit at the very beginning that one does well to be on one's guard with the people with whom one does not care to be silent,—but one does not go camping with those people,—or, as the case may be, if we, ourselves, have a guilty conscience or an empty head much talking serves its ends. And there is another situation in which it seems almost impossible to be silent. There is someone for whom we have cared very much. Things have changed, there has been a misunderstanding, we have altered or someone else has made trouble between us. And the first thing we notice is that we no longer dare to be silent together. Speech must be made to cover up our common lack of sympathy. We talk, how we talk,—anything, everything! Even when we are happy we run to places where there is no silence, but now, if only we can be as noisy as children and avoid the truth of the sad thing which has happened to us!

Again, let us admit at once that there are different kinds of silence: There is a bitter silence which is the silence of hate, and another which is that of evil thoughts, and a hostile silence, and a silence which may mean the beginning of a storm or a fierce warfare. But the only silence worth having is friendly and it is of that we need to think, and it is that we can have by the camp fire in our wilderness life.

Isn't it true after all that the question which most of us ought to ask ourselves seriously is not how many times we have talked but how many times we have been silent. Sometimes one wonders whether we are ever still and whether if we are to be silent, it is not a lesson which must be learned all over again. How many times have we talked in a single day? We can't tell, for the number of times is so great that we can't count them. And the times we have been silent? And I don't mean how many times we have said nothing. To say nothing is not necessarily to be silent. Well, we can't count the times we have been silent either, but that is because we haven't been still at all. Yet there is a big life in which there is no speech and no need of it. Are we never to give ourselves a chance to live that?

Do you remember your first great silence? Was it going away from someone you loved? Perhaps it was a joyous visit to your grandmother or to an aunt or to see a friend, but it meant leaving your mother and you had never left her before. Or maybe it was your first year at boarding school or your freshman year at college. Do you remember the silence that came over you then and all that filled it? And do you remember how it wore away but gradually—that grip the stillness had within you and upon you? You know now that that first silence will never be forgotten. Or was it a return to those you loved and you realized as never before how incomparably dear these people were to you and that only silence could express that dearness? Or was it the silence of a crowd—awe inspiring silence which foretells the acclaim of some great event of happiness or a cry of woe? Or the silence of the wilderness as you looked down from a mountain side into some great valley of lakes? Or was it the death of someone you loved, and the silence that overcame you forced you not only to suffer as never before but also to think as you have never done about the meaning of life?

In that first great silence how many things that are precious revealed themselves to us. There was love; we did not realize how it was woven into every fibre of our lives; there was companionship; we did not realize how bitterly hard it would be to forego it; there was new experience; till it came we could not have known how much a part of our lives the old experience was. How many things in us that had been asleep were suddenly awakened! How much was that great silence worth to us then and now? Perhaps an unhappy or stricken silence we called it then; but even if it meant death or separation was it after all completely unhappy? Have we taken into account the wealth of conviction, of deepened experience, of increased love it brought us? Could anything so rich be in any true sense unhappy?

"Silence, the Great Empire of Silence," cried Carlyle, "higher than the stars, deeper than the Kingdom of Death." The world needs silent men but even more, I think, does it need silent women. Carlyle—and you should get what you can of his books and read them—calls silent men the salt of the earth. Might not silent women or silent girls be called double salt? He says that the world without such men is like a tree without roots. To such a tree there will be no leaves and no shade; to such a tree there will be no growth; a tree without roots cannot hold the moisture that is in the earth and it will soon fade, soon dry up and let everything else around it dry up, too.

Have you not heard women and girls with an incessant silly giggle or a titter or a laugh that meant just nothing at all and yet which was heard, like the dry rattle of the locust, morning, noon and night? Nervousness partially; empty-headedness maybe, or a mistaken idea of what is attractive. Silliness of that kind has no place in camp. Nothing is more wearying, more lacking in self-control than such a manner, nothing so exhausts other people. Such giggling or laughing or silly talking is to the mind what St. Vitus's dance is to the body—an affliction to be endured perhaps but certainly not an attraction and not to be cultivated.

Is it not silence that opens the door to our best work? How about that work you enjoyed so much and did so well? How did you prepare for that? Yes, I know all about the work you bluffed through and even managed to get a high record in, but that work you really enjoyed, how was that done? Is it not silence, too, that opens the door to our dearest and deepest companionships, our profoundest sorrows, our greatest joys? Anyway this wilderness silence is all worth while thinking about, is it not?

Why should this great silence, this friendly wilderness power be considered anti-social? Really, is it not most social? Does it not bring us all nearer together, sometimes even when we are afraid to be nearer to one another? Does it not make us all equal, making us aware of those profound things in life which we all have in common? Silence can say, can teach, what speech can never, to the end of the world, learn to express. It is safe to say that as soon as most lips are silent, then and then only do the thoughts and the soul begin to live, to grow, to become something of what they are destined to be, for as Maeterlinck says, silence ripens the fruits of the soul. Never think that it is unsociable people or people who don't know how to talk who set such a value on silence. No, it is those who are able to talk best and most deeply, think best and most deeply, who, following the long trail, recognize the fact that words can never after all express those truths which are among us—no, neither love, nor death, nor any great joy, nor destiny can ever be expressed by word of mouth, by speech.

———

CHAPTER XVII
HOMEMADE CAMPING

It was our second day in camp,—a camp on the edge of the Maine wilderness. Around us were many lakes—ponds as the natives call them—Moosehead, Upper Wilson, Lower Wilson, Little Wilson, Trout Pond, Horse-shoe Pond, and a dozen others. About us on all sides were the forest-covered mountains, and burning fiercely, twenty miles distant, a large forest fire which filled the horizon with dense, yellow smoke.

From our camp, consisting of a red shanty, a log cabin in which I am now sitting, my dog beside me, thinking what I shall say to you about a remarkable family I saw, and, looking up at the cabin ceiling, its log ridge-pole and supports between which are birch bark cuts of trout and salmon caught in the lakes, of which I have spoken—from our camp we look out and down on a wonderful view. Immediately in front of the log cabin is a meadow, the last on the edge of this wilderness, then the serrated line of pointed firs, which marks the edge of the woods at the foot of the meadow. Beyond this line miles of tree-tops, pines, birches, maples, beeches, after that the shining lakes, and beyond them the mountains. There is not a house in sight. For that matter there *is* no house to be seen, not even a log cabin.

As was said, there is a meadow in front of the cabin, and over to the right beyond our view are two other meadows. In Maine—as far north as this, anyway—the farmers have only one crop of hay, and, when there is so much forest, and the winter is long, and cattle are to be fed, every meadow has to be counted upon for all it will bear of hay. It was a foregone conclusion that somebody would need and use the crop from the meadow down upon which my cabin looked.

And, sure enough, the second day we were in camp, along the road bumping and thumping over the big stones came a large hay wagon: behind it, rattling and jarring, a mowing machine and hay rake. But that hay wagon, what didn't it hold? In the first place, there was the driver, then a big packing box, a tent rolled up, sacks of feed for the horses, a baby's perambulator, three children, a woman, a hammock, a long bench, some chairs, including a rocking chair, and several small boxes, packed to overflowing with articles of various kinds. For an instant it looked as if they were house-moving, and then, recollecting that there was no house to which to move, I came to the conclusion that they were merely haying.

I watched them spread the big tent-fly and make it fast. I saw them take out the large packing box, converting that into a table, on which some of the children put flowers in an old bottle; I watched them set out the bench and chairs, swing the hammock, lay the improvised table with the enamel dishes which they took from the little boxes, and, in general, make themselves comfortable.

The children had pails for berries, and they began to pick berries in a business-like fashion. The woman sat in the hammock and took care of the baby—oh, I forgot to mention the baby. The farmer and his lad hitched and unhitched the horses, starting within a few minutes to work with the mowing machine, and leaving two of the horses tethered to a tree. Evidently this was work and a picnic combined—to me a new way of getting in your hay crop. But the more I watched it and thought about it the more I liked it. And their dinner with the berries as dessert—well, I knew just how good, there in the sunshine, with appetites sharpened by work, it must taste to them all.

Inside the cottage shanty of our camp, one member of the household, at least, had been doing her work in quite a different spirit. It seemed to me that there was nothing which this cook, a large, robust woman, with an arm with the strength of five, had not found fault with and made the worst of. Her first groan was heard in the morning at six o'clock—in getting up myself to go to my writing table I had cruelly awakened her—and, of course, as she went to bed only half after seven the night before, she had been robbed of her necessary sleep. As I say, I heard her first groan—the sun was shining gloriously, and I had already had a sun bath and a cold sponge and my morning exercises—while she continued to lie in bed and to make every subsequent groan until after seven o'clock fully audible.

She began that beautiful day and its work in resisting everything. She had never been in such a place before, and a very nice convenient camp we, ourselves, thought it. She groaned while she pumped water—I do not know whether she or the pump made the more noise. She complained loudly because of the mice. Oh, no, she could not set a mouse trap: she had never done such a thing before! And then, when we got a cat, she complained because of the noise the cat made in catching the mice. I do not know precisely what kind of a cat she expected, possibly a noiseless, rubber-tired cat, that would catch noiseless, rubber-tired mice. She would not carry water—even a two-quart pail full—her back was not strong enough. She had never seen such dishes as these we were using, nice, clean enamel ware dishes, with blue borders. She had never heard of such a thing as hanging milk and butter in a well to keep them cool. Dear me, she never even thought of going to such a place where they did not have ice that would automatically cool everything, and which the ice-man kindly handed to her in pieces just the size which she preferred. She said the spring—a beautiful spring whose waters are renowned for their purity and healthfulness much as the waters of Poland Spring are—she said that the spring had pollywogs in it and frogs. She could not string a clothes-line, but stood in tears near the big trunk of a balsam fir, holding the line helplessly in her hands and looking up to the branch not more than two inches above her head. While one of us flung the end of the clothes-line over the branch and made it fast to another she remarked with contempt, sniffing up her tears, that it was not a clothes-line, anyway, which was perfectly true, for it was only a boat cord, but it did quite as well. When she walked down from the meadow, that glorious golden meadow, where the happy family was picnicking and hay-making at the same time, and through which wound a little path down to the spring's edge, she lifted her skirts as if she were afraid they might be contaminated by the touch of that clean, sweet-smelling, long grass. Still groaning she would fetch about a quart of water. And groaning, still groaning, she went to bed at night "half-dead," as she expressed it, as the result of about five hours of work, in which she was all the time helped by somebody else.

Of course she was "half-dead." It is a wonder to me now, as I think of it, that she did not die altogether. Instead of taking things as they were in the sun-filled day, with its keen, crisp air, its wonderful view, instead of feeling something of the beauty and health and sun and wind-swept cleanness of it all, she had resisted every detail of the day, every part of her work, she had, in short, found fault with everything. This day, that would have seemed so joyous to some people, had not meant to her an opportunity to make the best of things and to be grateful for the long sleep, the sunshine, the invigorating air, the beauty, the light work, but merely a chance to make the worst of things, to throw herself against every demand made upon her.

Out in front of the cabin the farmer swept round and round with his mowing machine, his big, glossy horses glistening in the sunshine, the sharp teeth of the machine laying the grass in a wide swath behind him. He seemed peaceful and contented, although it was warm out in the direct sunlight, and the brakes were heavy and the horses needed constant guiding. Down below, nearer the spring, his wife swung in the hammock, and the children picked berries, fetched water, and were gleefully busy. It was a scene of simple contentment with life.

When the father came back for his dinner, which was eaten under the spread of a tent-fly and from the top of a packing box, decorated in the center with flowers and around the edges by contented faces, I said to him: "You seem to be having a jolly time."

"Why, yes, so we are," was his reply. "I offered the folks who own this meadow such a small sum of money for the hay crop I didn't think I'd get it. I thought some one else was sure to offer them more, but I guess they didn't, for I got it. You see, it's pretty far away from my farm to come out here haying."

"And so you make a picnic of it?"

"Yes, we are making a picnic of it. The children like it. It's great fun for them, and it gives my wife, who isn't very strong, a chance to rest and be out of doors. I enjoy it, too. I like to see them have a good time."

"Well," I said, before I realized I was taking him into my confidence, "I wish you could make our camp cook see your point of view."

"Why, don't she like it?" he asked innocently.

"Like it? I am afraid she doesn't. The other day it rained and leaked in through the kitchen roof onto her ironing board, and when we found her she had her head on the board and was crying."

"Well, that's too bad," he said. "Why didn't she take that board out of the way of the leak? We don't mind a little thing like a leak around here, especially when folks are camping. Having her feel that way must make a difference in your pleasure. Well, there is ways of taking work. Now, probably, she's throwing herself against her work, and making it harder all the time."

"That's exactly what she is doing," I commented dryly.

"It's a pity." There was sympathy in his voice. "For it's such a lot easier to make a picnic out of what you are doing—homemade camping, we call this. My folks always feel that way about it. Even the hardest work is easier for taking it the right end to. My children are growing up to think, what it doesn't hurt any man to think, that work is the best fun, after all. It's the only thing you never get tired of, for there is always something more to do."

———

CHAPTER XVIII
THE CANOE AND FISHING

It was my somewhat tempered good fortune, several years ago, to spend two or three weeks in an exceedingly bleak place on a far northern coast. The only genial element about this barren spot was its sea captains, and whence they drew their geniality heaven only knows. They made me think of nothing so much as of the warm lichen which sometimes flourishes upon cold rocks. There strayed into this neighborhood a couple of canoes. "Waal," exclaimed one of the old salts, viewing this water craft skeptically, "it's the nearest next to nothing of anything I have ever heard tell on."

And that is precisely what the canoe is: the nearest next to nothing in water craft which you can imagine. It is in precisely this nothingness that its charm lies, its lightness, its grace, its friskiness, its strength, its motion, its adaptability to circumstances. There are times when it acts like a demon, and there are other times when its intelligence is almost uncanny. The canoe is always high spirited, and, with high-spirited things, whether they be horseflesh or canoe, it does not do to trifle. The girl who expects to take liberties with the canoe has some dreadful, if not fatal, experiences ahead of her. Several years ago I was out in a motor boat with some friends. Two of them had been, or were, connected with the United States Navy; another was my sister, and a fourth was a college friend. My friend happened to see a pistol lying on a seat near her. She had never had anything to do with pistols, and, on some insane impulse of the moment, she picked it up and leveled it at me. I was stunned, but not so the men on the boat. Such a shout of rage and indignation, such a leap to seize the pistol, and such a rebuke, I have never been witness to before. These men were navy men, and they knew how criminally foolish it is to fool with what may bring disaster. It is those who know the canoe best and are best able to handle it, who are most cautious in its use. Those of you who expect to treat it as you might the family horse would do well to look out.

The canvas-covered cedar canoe is the best. If you are going to take a lot of duffle with you, the canoes will have to be longer than you need otherwise have them: about eighteen feet, and only two people to a canoe. The canoe will cost you from twenty-five dollars up, and this item does not include the paddle. The paddle should be bought exactly your own height; it will then be an ideal length for paddling. Its cost will be a little more or a little less than a dollar and a half. You should have a large sponge, tied to a string, on one of the thwarts. This you will use for bailing when necessary.

If you have had any experience with a canoe, you will not abuse it, and will not need to be told not to abuse it. If it is a light one, and you are a strong girl, you should learn to carry it Micmac fashion on the paddle blades, a sweater over your shoulders to serve as cushion. Watch a woodsman and see the way he handles a canoe. One of the very first things you will observe is that he never drags it about, but lifts it clean off the ground by the thwarts, holding the concave side toward him. Also, you should observe his soft-footed movements when he is stepping into a canoe. If a canoe is not in use it should be turned upside down. Never neglect your canoe, for a small puncture in it is like the proverbial small hole in a dike. If you let it go, you will have a heavy, water-soaked craft or a swamped one. Water soaking turns a seemingly intelligent, high-spirited canoe, capable of answering to your least wish or touch, into the most lunk-headed thing imaginable, a thing so stupid and so dead and so obstinate, that life with it becomes a burden. Remember that the wounds in your canoe need quite as much attention as your own would.

The balance of a canoe is a ticklish thing. To the novice, the day when she can paddle through stiff water while she trolls with a rod under her knee and lands a two- or three-pound salmon unaided, seems far off. I am by no means a past-master in the art of canoeing, yet I have often done this, and am no longer troubled by the question of balance in a canoe. So much for encouragement! Most of an art lies, granting the initial gift for it, in custom or habit. Make yourself familiar with the traits of your canoe, work hard to learn everything you should know about it, and your lesson will soon be learned.

When you are going to get into it, have your canoe securely beside a landing, and then step carefully into the center and middle. Bring the second foot after the first only when you are sure that you have your balance. The next thing is to sit down. Be certain that it is not in the water. The only satisfactory recipe for this delicate act is to do it. No girl should step into a canoe for the first time without some one at the bow to steady it. Very quickly you will learn clever ways of using your paddle to help in keeping the balance. Until you do, you can't be too careful, or too careful that others should be careful. Take no chances in a canoe. If any are taken for you, hang on to your paddle. It is well to have an inflatable life-preserver, but, best of all, is it to know how to swim. Never move around in a canoe, or turn quickly to look over your shoulder. A canoe is a long-suffering thing, but once "riled" and its mind made up to capsize, heaven and earth cannot prevent that consummation and your ducking or even drowning.

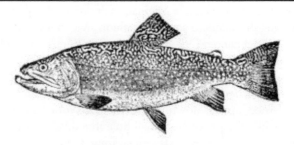

BROOK TROUT

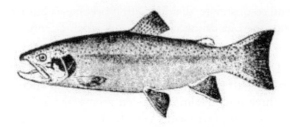

RAINBOW TROUT

SMALL-MOUTH BASS

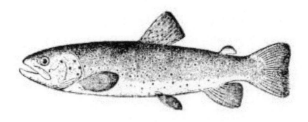

BROWN TROUT

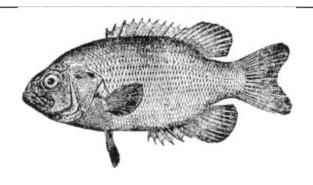

ROCK-BASS

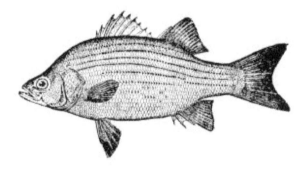

WHITE BASS

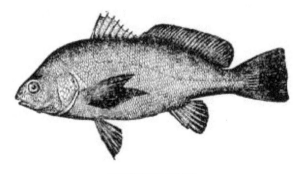

SHEEPSHEAD

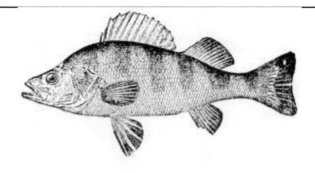

YELLOW PERCH

PIKE

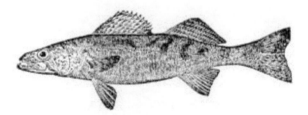

PIKE PERCH

PICKEREL

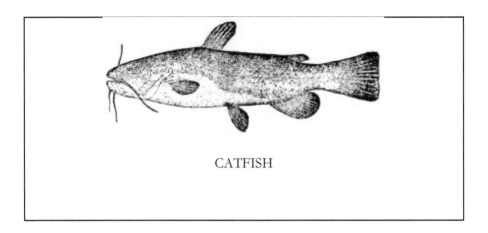

CATFISH

Become skillful in the use of the paddle, and the best way to learn is through some one who knows how. Paddling is an art and a very delightful one, requiring much skill of touch and strength. Although as a girl I cared most for rowing, I have in the last ten years become so devoted to the paddle stroke, to its motion and touch and efficiency, that rowing only bores me. Get some one, a brother, a father, a friend, a guide, to teach you the rudiments of paddling. These once learned, canoeing is as safe as bicycling and not more difficult. It is all in learning how.

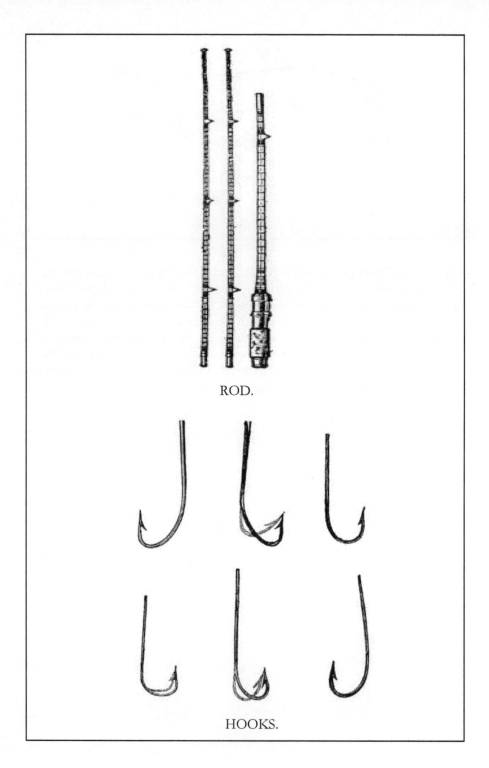

ROD.

HOOKS.

SIMPLE WINCH REEL.

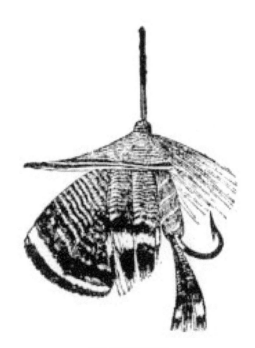

TROUT FLY.

TROLLING SPOONS.

The writer is an old-fashioned fisherwoman and goes light with tackle. However, I have noticed that the simplicity of fishing tackle does not in the least interfere with luck. If you are going to fish with worm, hook, and sinker, you will need no advice. Perch, pickerel, black bass, cat-fish, and others to be caught in still fishing, will be your quarry. As a rule you will troll for pickerel and pike, and there is no sport more pleasant in the world than that which is to be had at the end of a trolling spoon: the motion of the boat, the vibration of the line, the spinning of the spoon, and then the sudden strike, with all its possibilities for taking in big fish. I defy anyone to have a more exciting time than netting a salmon from a trolling line and landing it successfully in a canoe. But this is not a thing to be attempted by the novice. Much better let the salmon go and save yourself a ducking.

The finest art of all fishing is fly-fishing. One either does or does not take to it naturally, after one has been taught something of the art by brother, father, or guide. Alas, that the fish greediness of campers is making good fly-fishing, even in the wilderness, more and more difficult to get! Personally, if I am after trout or salmon, "plugging" or "bating," as it is called, seems to me an unpardonably coarse and stupid sport. Yet our lakes have been so abused by this process that fly-fishing is frequently impossible. To sit or stand in a canoe, casting your line, the canoe taking every flex of your wrist; to see the bright flies, Parmachenee Belle or Silver Doctor—or whatever fly suits that part of the country in which you are camping—alight on the surface as if gifted with veritable life, and then to be conscious of the rush, the strike, and to see a rainbow trout whirling off with your silken line, is to experience an incomparable pleasure. To have a strike while the twilight is coming on, a big fellow, with the line spinning off your reel as if it would never stop, to see your salmon leap into the air and strike the water, to reel him in, then plunge! and down, down he goes; to feel the twilight deepening as you try to get him in closer to the canoe again; to know suddenly that it is dark and that the hours are going by; to feel your wrist aching, your body tense with excitement; to think that you are just tiring him out, that you have almost got him—almost, then a rush, a plunge, the line slackens in your hand, and he is gone. That is fisherman's luck, and great luck it is, even when the fish is lost.

ROD CASE.

FELT-LINED LEADER BOX.

CASE FOR TACKLE.

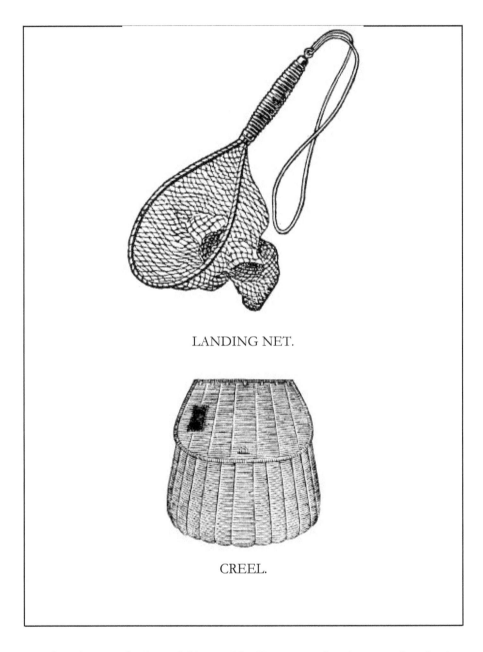

LANDING NET.

CREEL.

Only a few words about fishing tackle. Have a good rod or two, but don't begin your experience at fishing with expensive tackle. The cheaper rod will do quite as well until you learn what you want. For trolling the best rod is a short steel one. For fly-fishing you will always use split bamboo or some similar wood. You will have accidents, so have reserve tackle to fall back

upon.   In   any   event   do   not   buy   a   heavy
rod, and never buy anything with a steel core in it. If you can afford it, get a
first-class reel, one that works easily and is of simple mechanism. A simple
winch reel is the best. Avoid patented contraptions. While you are using them
hang your rods up by the tips. In any event keep them dry and in as good
condition as possible. Enameled silk line you must have for all trout fishing.
For other kinds of fishing it does not so much matter what you do use,
provided the line is strong and durable. Be sure to have extra lines to fall back
on.

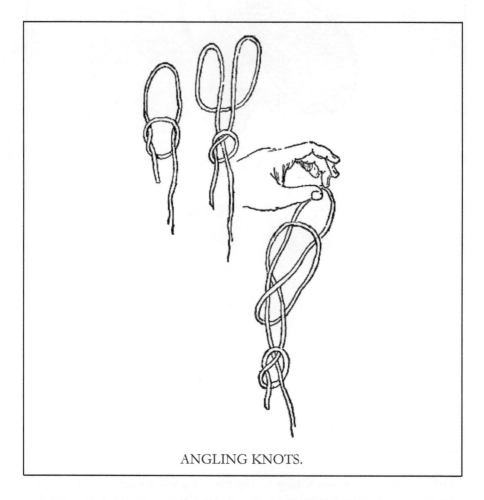

ANGLING KNOTS.

Leaders, the details about flies to be used, their color, angling knots made in
fastening leaders or line or fly, methods for keeping your flies in good order
and condition, the use of the landing net, necessary repairs to be made, the
skill of the wrist in casting, the best sort of trolling, the care of fish, all these

things will come to you through experience, and all suggest how much, how delightfully
much, there is to be learned in the best of all sports.

Go to some first-rate sporting goods' house for your flies; they will tell you what kinds you need, as well as answer other questions.

———————

CHAPTER XIX
THE TRAIL

A girl who has learned to camp will not only have her own pleasures greatly increased, but she will also add to those of her friends, becoming a better companion for her chums, her father, her brother; for camping, if it is anything, is a social art. It is far better for a girl to be out in the world which demands all of one's attention, one's eyes and ears and nose and feet and hands and every muscle of the entire body, than to be leading a sedentary life at home, or analyzing emotions or sentimentalizing about things not worth while. The big moose which unexpectedly plunges by provides enough emotions to last a long time; the land-locked salmon that threatens to snap the silken line, enough excitement.

You can't learn all that there is to be learned in the school of the woods through one camping expedition. It would be rather poor sport if you could. Don't be afraid to ask questions about what you don't know. Keep on asking them until you are wood-cultivated. The wilderness is your opportunity to make up for those vitally interesting facts about life which are not taught in schools. Above all, have a map of the country in which you are, and study it. Keep that map by you as if it were Fidus Achates himself, and refer to it whenever there is need. The girl or woman in camp who never knows where she is is a bore, sponging upon the good-nature and intelligence of others who have taken the trouble to familiarize themselves with the lie of the land. Such a girl never makes any plans, never takes the initiative, never gives anyone a sense of rest from responsibility. There are girls and older women who think it rather clever to be unable to tell east from west, north from south. I may say here that in camp they belong to the same class of foolish incompetents who in college boast that they cannot spell—presumably because they are devoting themselves to a much higher call upon their intelligence than anything so superficial as spelling! If camping means anything in the world, it means coöperation, and this coöperation should be all along the line.

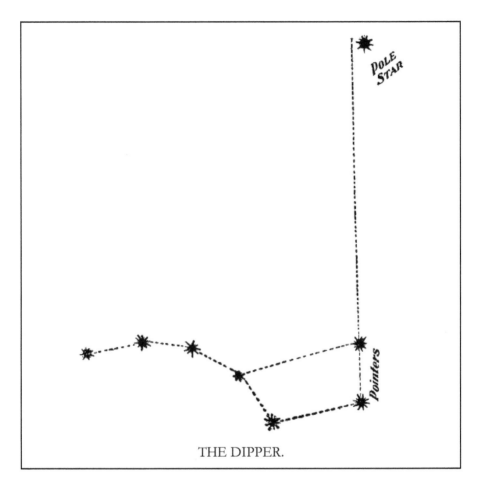

THE DIPPER.

If you have an innate sense of direction, train it. If you have none, do not venture out into the wilderness except with someone who has. Always tell people where you are going. If you are not familiar with the use of a rifle you would better have a shrill whistle or a tin horn to use in case you want to summon anyone. Sun and wind should be part of your compass; the trees, too. You will, of course, learn how to blaze a trail, and the sooner you do this the better, for it is good training in following out a point of the compass. The wilderness is full of signs of direction for your use, some of which are certain to be serviceable at different times, and some of which will not prove dependable. The sun rises in the east and sets in the west. At high noon of a September day, if you turn your back squarely to the sun, you will be looking directly north. The wind is a helper, too. When the sun rises, notice the direction of the wind, and, while it does not shift, it will prove a good compass or guide. If it is very light, wet the finger and hold it up. By doing this the wind will serve you as a compass. Remember, also, that the two lowest stars of the Big Dipper point toward the North Star, which is always a guide to be used in charting a wilderness way. Also on the north sides of trees there is greater thickness to the bark and more moss. This is, I suppose, because the trees, being unexposed to the sunlight on the north side, retain the moisture longer there. Some say, too, that the very topmost finger of an evergreen points toward the north. Even in civilization they usually do. To become familiar with a compass is a very simple matter. Every boy learns this lesson, and there is no reason why girls should not do the same. Never buy a cheap compass; it is not to be relied upon. To the amateur in the woods a good one is not a friend at which to scoff. A few expeditions out behind the cabin will teach you all you need to know about its use. If by some miscalculation a girl should get lost, let her realize then that the great demand is that she shall keep her head on her shoulders, where it has been placed, and where she will need to make use of it. Let her sit down and think, reviewing all that has happened, and trying to solve the problem of what she is to do. A panic is the last and worst thing in which she can afford to indulge. To most people at some time or other comes the conviction that they are lost—a conviction happily dispelled in nine hundred and ninety-nine cases out of a thousand. In this, as in everything, a miss is as good as a mile, and one does well to make light of unavoidable mistakes.

FAWN

DOE

BUCK

CARIBOU

MOOSE

If, by any chance, you should be lost, don't run around. If you have no compass or if darkness is coming on, settle down where you are. Devote your energies to occasional periods of shouting and to building a camp fire, keep your body warm and dry and your head cool. *You will be found.* And remember that there are no wild creatures to be feared in our camping wilderness. You have nothing of which to be afraid except your own lack of common sense. Here is a chance for your "nerve" to show itself.

RED SQUIRREL

FLYING SQUIRREL

GRAY SQUIRREL

RABBIT

AMERICAN SABLE

CHIPMUNK

WEASEL

BLACK BEAR

RACCOON

MINK

PORCUPINE

SKUNK

WOODCHUCK

RED FOX

As you go through the woods, cross the ponds and lakes, climb mountains, your luncheon in your pocket, compass and knife and cup and match-box all ready and friendly to your hand; as you feel the wilderness becoming more and more your empire, be sure that you do not abuse the privileges which are revealed to you. The more gentle and considerate you are in this life which has opened itself up to you, the more it will tell you its secrets. That you should leave disfiguration and destruction and bloodshed behind you does not prove that you are in any sense a true sport. The camera is one of the best guns for the wilderness. It is better to be film-thirsty than bloodthirsty. A girl who is in earnest about camera shooting can test her "nerves" quite sufficiently for all practical purposes. How about facing, or chasing, a six- or seven-hundred-pound moose, plunging down through a cut or a trail, and having the nerve to press the bulb at just the right moment? Or a big buck? Or a little bear? Or a porcupine? A good kodak and some rolls of film are all that is needed to begin the work of photography. A fine way to do, if you intend to go into the matter seriously, is to get some book on nature photography and make a thorough study of it. Other books, too, there are, which will be full of profit for you as you come to know the wilderness life. Begin with Thoreau, John Burroughs, John Muir, Stewart White, Ernest Seton Thompson, and these will lead you on and out through a host of nature books and finally into a more technical literature on hunting, camping, and the wilderness life in general.

I believe that in the end an intelligent study of the woods made with eyes and ears, heart and mind, notebook and book, will bring down more game than any shotgun or rifle ever manufactured. I have seen guide-books of northern wildernesses whose collective illustration suggested only the interior of some local slaughter house. No tenderfoot myself, for, when the first shotgun was placed against my shoulder, I was so little that its kick knocked me over, I do not write this way because I am unfamiliar with the pleasures of well-earned or necessary game, but because I have tried both ways and I prefer a friendly life in the wilderness. To kill what you see, just because you do see it, to set big fires, to be wasteful, to take risks in your adventures, are no signs that you know the woods—and they are most certainly no guarantee of your love.

———

CHAPTER XX
CAMP DON'TS

Don't forget your check list.

Do make your plans early for the camping expedition.

Don't be dowdy in the woods. Dress appropriately.

Do keep a clean camp. Otherwise you will go in for hedgehogs, skunks, flies, and other disease-breeding pests.

If in doubt about drinking water, don't drink it—at least, not till it is thoroughly boiled.

Do be independent. Camp is no place for necklaces, however beautiful.

Don't start out camping with a new pair of shoes on your feet.

Do keep from adding to the things you want to take with you, or you won't be able to reach the "jumping off" place.

Don't forget your fly "dope."

If your appetite is good, be polite to the cook.

Don't forget the box of matches.

Don't be foolhardy. It might take too long to find you. If you feel that way, have somebody attach a tump line to you.

If you have an open stove, when you go off for the day, be sure to close it.

Don't be afraid to ask questions—everybody does.

Do help others with the work.

Don't cut your foot with the axe. It will not add to the pleasures of camp life.

Dish-washing is not pleasant work. Do your share just the same.

Don't step on the gunwale of the canoe, and upset it, or trip over a thwart. The canoe is a ticklish craft.

Do conform to the camp routine. Don't keep the dinner waiting, delay the fishing expedition, or call out a search party.

Don't be ignorant of the topography of the region in which you camp. By not studying the map for yourself, you will give others a lot of trouble.

Listen to what your guide says.

Remember, I shall be glad to answer brief, pointed questions, addressed to me at

CAMP RUNWAY,
Moosehead Lake, Greenville, Maine.

THE END